The Bodies of Women

Rosalyn Diprose argues that injustice against women begins in the ways in which social assumptions about sexual difference constitute women's embodied existence as improper and secondary in relation to men. *The Bodies of Women* intervenes into and brings together debates about sexual difference, ethics, philosophies of the body and theories of self from the continental philosophical tradition to show that the usual approach to ethics both perpetuates and remains blind to these mechanisms of the social subordination of women.

Crucial to this argument is the belief that such injustice is revealed through critical analyses of discourses which regulate sexual difference: from ethics itself (contractarian ethics, ethics of care and some feminist interventions into the ethics of reproductive practices) to those discourses of 'the body' which purport to merely describe, rather than constitute and regulate, embodied existence.

While these critiques draw on insights from Anglophone feminist theory and the continental philosophical tradition, these are also subject to scrutiny. Consequently the book includes critical rereadings of Hegel, Nietzsche, Merleau-Ponty, Derrida and Foucault as well as productive engagement with contemporary feminist scholars such as Irigaray, Cornell and Young. What emerges is a unique approach to the ethics of sexual difference which both locates and subverts mechanisms of sexual discrimination.

Rosalyn Diprose is a Lecturer in philosophy at the University of New South Wales. She is the co-editor, with Robyn Ferrell, of *Cartographies: Poststructuralism and the Mapping of Bodies and Spaces*.

The Bodies of Women

Ethics, embodiment and sexual difference

Rosalyn Diprose

ROUTLEDGE

London and New York

First published 1994
by Routledge
2 Park Square, Milton Park, Abingdon, Oxon, OX14 4RN

Simultaneously published in the USA and Canada
by Routledge
270 Madison Ave, New York NY 10016

Transferred to Digital Printing 2010

Typeset in Palatino by Intype Ltd, London

British Library Cataloguing in Publication Data
A catalogue record for this book is available from the British Library.
ISBN 0-415-09782-7 (h/bk) ISBN 0-415-09783-5 (p/bk)

Library of Congress Cataloging in Publication Data
Diprose, Rosalyn.
The bodies of women : ethics, embodiment, and sexual difference / Rosalyn Diprose.
p. cm.
Includes bibliographical references and index.
ISBN 0-415-09782-7 – ISBN 0-415-09783-5 (pbk.)
1. Feminist ethics. 2. Woman (Philosophy)–Moral and ethical aspects. 3. Human reproduction–Moral and ethical aspects. 4. Sex differences–Moral and ethical aspects. 5. Body image. 6. Body, Human–Moral and ethical aspects. 7. Feminist theory. I. Title.
BJ1395.D57 1994
176'.082–dc20
93-37969
CIP

Publisher's Note
The publisher has gone to great lengths to ensure the quality of this reprint but points out that some imperfections in the original may be apparent.

Contents

Introduction

For some time, ethics, as a branch of Anglophone philosophy, has tended to focus on the nature of moral judgement (to secure its rational basis) or on the nature of the moral principles which do or should govern social relations (to secure their universal status). Behind this inquiry lies the conviction that a moral code can and should maintain our social order, protecting it against transgression and disintegration. At the same time there has been an increasing belief that, while we may need a moral code for this purpose, morality is inadequate to its task (Poole 1991) and is often oppressive in practice (Macintyre 1985; Benhabib 1992). This book adds to these voices of discontent but from a particular perspective. It is based on the related convictions that, insofar as a moral code may maintain some semblance of order, this is at the expense of justice for women and that the usual approaches to ethics perpetuate and/or remain blind to such miscarriages of justice. My general aim then is to develop an approach to ethics which takes sexual difference into account.

The new feminist ethics do just that: they attempt to take sexual difference into account. Chapter 1 discusses some of these: the ethics of care (Gilligan 1982) and its derivatives. What these challenge, in various ways, is the abstract individualism of ethics, the validity of general moral principles and the assumed sexual neutrality of moral judgement. With this challenge comes the insistence that, when assessing the moral worth of a particular activity, when regulating social relations or when engaged with a particular other, we need to take differences into account. In particular we need to acknowledge that the social context is patriarchal and that women have different experiences from men within this context. While a welcome intervention into ethics, I am left doubting whether these approaches can save or give voice to sexual and other differences. My doubt arises from the absence of a convincing account of how sexual difference is produced and maintained within patriarchal social relations. Without this I cannot share the confidence, apparent in these feminist ethics, that women's different experiences can be articulated and thereby saved from injustice by the means suggested (dialogue with the other and related practices such as community action and giving attention to context).

That we live in complex and difficult times which call for a revolution in ethics (Irigaray 1991), and a well developed sense of irony, is illustrated by the following, somewhat bizarre, event. In Australia in June 1993 Mr Damian Taylor won the Miss Wintersun Quest. Not only did he win the 'charity queen' category, but he took out the overall title, allowing him to go on to contest the Miss Australia Awards. The question on my lips, as this was announced on my radio, was echoed by an exasperated woman journalist. 'On what basis could you possibly award a women's beauty contest to a man?' she asked of one of the female judges. 'Not on the basis of beauty (we don't call them beauty contests any more),' the judge replied, apparently missing the irony of the question, 'but on the basis of other criteria such as intelligence, interpersonal skills, manners, grooming, comportment and body shape.'

In the wake of *The Man of Reason* (Lloyd 1984), we should no longer be surprised when 'intelligence' or rationality is automatically equated with masculinity. But the idea that a man could win a woman's 'beauty' contest on criteria concerning bodily comportment and shape really did leave me gasping. Yet, I thought to myself, perhaps this is not surprising either. The literature on transsexualism, for example, is abundant with observations of how male to female transsexuals perform (male) ideals of feminine bodily comportment better than women (Stone 1991). Parody or a deliberate masquerade could explain Damian Taylor's victory. As I eagerly turned on the television, any hopes of critical irony along these lines were dashed. It turned out that Damian is as 'straight' as they come, the epitome of modern manhood: a lifesaver and naturopathy student with proud girlfriend and parents at his side. Sincere new-sensitive-man meets average Australian bloke and beats beautiful Australian women on their own terms.

While it is difficult not to feel incensed by all of this, it could be read as a minor victory for feminist ethics. After all, many feminists have vigorously opposed beauty contests, and any other practices, which objectify women's bodies and/or endorse male control over women's bodies. Admitting men into the objectification process would seem to displace women as a target and make men the object of sexual subjection. But if this can be construed as a victory for feminist ethics, it is a hollow one: the only losers are women. One kind of injustice is merely displaced by another in that it seems more appropriate to read this as a victory for men in the wake of equal opportunity. What makes this victory particularly disturbing is that it occurred in a context which should have favoured the women contestants insofar as their bodies signify womanhood. And what makes any injustice here difficult to articulate, locate and correct is the apparent absence of any male oppressor: the judgement was made and endorsed by women.

Perhaps I am making too much of this particular incident. After all, it did take place on the Gold Coast (half way between the Costa del Sol and Las Vegas) where peculiar things are known to happen. And it was only a 'beauty' contest, hardly of consequence to the vast majority of women who avoid them. Yet, I find it difficult to dismiss the incident as isolated and inconsequential. If we transport Damian Taylor and the other female contestants into the job market or the law courts, the ethical import of Damian's victory becomes more serious. If a man can be judged more meritorious on criteria developed specifically for, and therefore presumably favourable to, women's modes of embodiment, what does this say about the evaluation of sexual difference in general?

This book begins with the conviction that the Wintersun Quest is not an isolated case. It merely makes explicit what is obscured elsewhere: that the evaluation and regulation of sexual difference in patriarchal social relations is implicitly based on the valorisation of a particular kind of male body (Gatens 1991b). Added to this is the conjecture that the moral, legal, industrial and interpersonal evaluation of sexual difference is productive: it produces the modes of sexed embodiment it regulates. Further, a key argument throughout this book is that any injustice experienced by women begins from this mode of production and maintenance of sexual difference. Finally, while this process of production has no particular author to arrest and put on trial, we can challenge the process itself. Any revolution in ethics begins with the ethics of ethics: a cross examination of universalist ethics and other discourses complicit in the evaluation and regulation of sexual difference. It is my belief that current aspirations for opening a fairer deal for women and a wider range of possibilities for living must at least be accompanied by this kind of (ongoing) interrogation.

As I assume few of us have participated in beauty contests, at least not willingly, the elaboration of these claims begins in chapter 1 with a discussion of other kinds of contests: those involving the maternal body. By examining aspects of the debates over the ethics of reproductive practices, particularly surrogacy, the aim is to locate in what ways and for what reasons conventional ethics, contractarian ethics in particular, cannot fairly accommodate women's specific modes of embodiment. The discussion reveals a problem with the concept of the individual assumed in ethics (where the individual is said to be self-contained and the original owner of property in their body) and a problem with its model of social exchange (where the male body is assumed to be the only valid currency). With these assumptions in place, the ethicist inadvertently discriminates against women in the name of justice and preservation of the 'common good'. Also revealed is a tendency, in feminist interventions into the ethics of reproductive practices, to take over these assumptions and perpetuate the exclusions they effect. The critical appropriations of ethics of care, also discussed here, avoid some of these difficulties through their critiques of the contract models of the individual and social relations. However, as mentioned, I find a problem in the means suggested there for accommodating sexual and other differences (a problem revisited briefly in chapter 2 and more directly in chapter 3).

Chapter 2 picks up on two insights which emerge from, but remain undeveloped in, feminist critiques of universalist ethics: the claim that social identity is sexed and embodied and the idea that identity is socially constituted in relation to others. The suggestion developed in this chapter, through a discussion of the work of Michel Foucault and Luce Irigaray, is that, if universalist ethics regulates social exchange through which sexed identity and difference is produced and if this perpetuates injustice against women, then ethics needs to be understood as the problematic of the constitution of embodied, sexed identity. There are two parts to this problematic: how one's embodied ethos is constituted by social discourses and practices (including ethics) and how one's identity is constituted in relation to others. Foucault attends to the first part. His work on disciplinary power is useful for articulating the ways in which embodied identities are constituted, normalised and marginalised. And it is useful for explaining why injustice is difficult to locate and correct. The surveillance and moral regulation of the maternal body is reassessed here in terms of its consequence for the normalisation of the social body as a whole.

No ethics which is critical of mechanisms of subjection would be complete without a gesture towards other possibilities. Foucault's gesture lies in his aesthetics of self: recreating the self as a corporeal work of art, without reference to the disciplinary moral code. This ethics of difference is found to be problematic for a feminist ethics on the grounds that the male body is already considered a work of art in comparison to women's modes of embodiment (witness Damian Taylor). The suggestion made in this discussion is that the value and status enjoyed by men in patriarchal social relations is generated through the constitution of women's modes of embodied existence as other to the norm. This suggestion, explored through the work of Irigaray and other feminists, evokes the second part of the problematic I have called ethics: how sexed identity is constituted through the other. The issue for Irigaray, and for a feminist ethics in general, is the extent to which this production of sexed identity and difference involves the overvaluation or denegration of women's modes of being. It is this productive evaluation of women's embodied existence in terms of virtue or shame which provides the basis of women's exclusion from social exchange. What also emerges from this kind of analysis is a different model of agency and a different understanding of how change can occur from that of the liberal empiricist tradition. New possibilities for women are said to emerge from an interrogation of the usual modes of production of sexual difference. This idea is explained and put into practice in chapters 3 and 4.

The approach to ethics suggested in chapter 2 is developed throughout the rest of the book through critical re-readings of aspects of the French and German philosophical tradition with assistance from some Anglophone feminists. The re-readings are critical because, while this tradition is useful for elaborating the approach to ethics I am claiming is necessary for our epoch, it has its own problems in adequately addressing sexual difference.

Chapters 3 and 4 explore the idea that sexed identity and difference is constituted through, rather than prior to, social exchange. In chapter 3 Hegel is credited for taking differences seriously in this way. But he is also found guilty of legitimating the subordination of women. His ethics of communal unity is based on a limited understanding of sexual difference: woman is represented as man's complement and her difference is said to be transparent and reducible to man's identity. This discussion, while directly critical of Hegel, also serves as a warning against those contemporary versions of a dialogical or contextual ethics mentioned in chapter 1. While arguing that Hegel's ethics is normalising to the detriment of women, the analysis also demonstrates how, within Hegel's own terms, the production of sexed modes of embodiment is such that his aspirations for a universal ethos (and the subordination of women this involves) is undermined.

Chapter 4 deals with contemporary attempts to move beyond Hegel's limited model of sexed identity and difference (including those by Mauss, Derrida, Heidegger and a number of 'deconstructive' feminists). These reassessments of the productive dimension to the self/other relation suggest an ethics based, not on contract, dialogue or the unity of identity and difference, but on the 'gift'. From an account of the 'gift', which suggests the impossibility of autonomous identity, the discussion moves to an assessment of deconstruction as a means of opening other possibilities for women's existence beyond those which position women as 'other' to men. The suggestion throughout the chapter is that a feminist ethics needs to be wary of moving too quickly from an interrogation of the material effects of representations of sexual difference upon women's modes of embodiment to the claim that the gift of other possibilities for women has already arrived. The more hasty this move, the more likely the means of discrimination will be left in place.

With that caution in mind, chapters 5 and 6 return to a more direct analysis of the social constitution of embodied existence but with the added insight that sexed bodies are constituted within an economy of representation of sexual difference which limits possibilities for women. Chapter 5 explores this theme through a critical appropriation of Nietzsche's philosophy and chapter 6 through existential phenomenology. The general aim in both chapters is to demonstrate how injustice begins with the ways in which women's bodies are constituted as improper in relation to men by social discourses and practices. Chapter 6 also signals a return to consider the ethics of reproductive practices in the light of intervening analyses. The account is opened up into a more general discussion of biomedical ethics and the role of biomedical science (the ultimate discourse on the body) in the constitution and regulation of sexed embodiment.

While my hope is that each chapter stands by itself as a contribution to the rethinking of the ethics of sexual difference, each also builds on the themes which come before. Hence, some issues reappear for further critical attention. Under consistent review, for example, are conventional notions of moral agency, autonomy, justice and freedom and the concept of the individual upon which they depend: the self who governs and owns property in their body. Related to these is an ongoing critique of models of social relations based on atomised individualism including refutations of claims that women's subordination is based primarily on male control over women's bodies or that women's freedom can be secured by keeping our bodies to ourselves.

One final introductory comment: given the current emphasis on identity politics, it may seem a limitation to frame the discussion around the maternal body and the general distinction between men and women rather than considering other kinds of identities, sexualities and differences between women. As I hope will be obvious, it is in deference to, rather than at the expense of, such possible differences that this book is written. However, what should also be obvious is that I do not think any sexed identity escapes contamination by those dominant discourses which privilege heterosexuality and subsume women under a general category as man's other and as potential, actual or failed mothers. To make these discourses, rather than particular kinds of women, the object of ethics is to undermine, rather than repeat, their normalising, discriminatory and totalising effects. It is towards this end, the keeping open of determinations of women's existence beyond virtue and shame, that this book is directed.

Acknowledgements

The ideas which have taken shape in this book have been informed by many encounters. I am especially indebted to Cathy Vasseleu, Robyn Ferrell and Jennifer Biddle whose own work, intellectual generosity and friendship have stimulated and sustained me. My thanks also to Genevieve Lloyd, Moira Gatens, Paul Patton, Marion Tapper, Paul Komesaroff and Liz Grosz for inspiration, encouragement and/or critical attention to my work. Students in courses I have convened at the University of Sydney, the University of New South Wales and the Flinders University of South Australia have taught me much and for that I am grateful. I would also like to thank Adrian Driscoll, Leigh Wilson, Rebecca Johnson and the anonymous readers at Routledge for their advice and support. Barbara Baird assisted with the preparation of the manuscript and has my gratitude for this, for her friendship and for a number of insights informing my work. And a special thanks to Liz Diprose and the rest of my family for all those conversations which have shaped my ideas in incalculable ways.

I have drawn on material from several published papers: 'The body biomedical ethics forgets', in Paul Komesaroff (ed.) *Medical Ethics and the Body: Postmodern Perspectives*, Sydney: Allen & Unwin (forthcoming); 'Nietzsche and the pathos of distance', in Paul Patton (ed.) *Nietzsche, Feminism and Political Theory*, London: Routledge (1993); 'Foucault, Derrida and the ethics of sexual difference', *Social Semiotics*, 1: 2 (1991); 'In excess: the body and the habit of sexual difference', *Hypatia*, 6: 3 (1991). Part of my initial research was conducted in the History of Consciousness program at the University of California, Santa Cruz in 1988/89 with the generous assistance of a fellowship from the American Association of University Women. I am grateful to both programmes for their support.

Chapter 1

Feminism and the ethics of reproduction

Ethics, embodiment and sexual difference are three themes which need to be thought together. That they rarely are in practice presents certain difficulties for both ethics and feminism. In order to highlight these difficulties and to set the scene for further analysis, I will begin with discussion of a particular kind of body, a pregnant body, and a particular ethical problem, surrogacy. What one encounters in such an excursion into the field of biomedical ethics is an inability to deal with specific kinds of bodies. Pregnant bodies, for example, continue to pose a problem for biomedical ethics, particularly within deliberations on the ethics of abortion, surrogacy and the use of reproductive technology. This inability to account for a mode of sexual difference, I take to be exemplary of ethics in general.

One rarely finds mention of the body in discussions of the ethics of biomedicine, despite what would seem to be an obvious point: that the object of biomedical theory and practice is the body. It is assumed that biomedical ethics regulates, not so much relations between bodies (discursive and human), but relations between self-present, autonomous, disembodied individuals. And it is thought that these relations take the form of contracts (implicit or explicit) which can be governed by universal ethical principles. Yet, despite the apparent insignificance of embodiment, if the spectre of a pregnant body is raised, this model of social relations and its universal principles are found to be wanting.

Surrogacy, for example, has been practised without much fuss for as long as written memory. But the ruling in the case of Baby M in New Jersey in 1987 changed all that as a consequence of bringing surrogacy within the contractarian paradigm of social relations. Judge Sorkow, under the umbrella of contract law, upheld the surrogacy contract against the surrogate mother's wishes and ruled in favour of the genetic father's paternity rights. While this ruling sparked widespread outrage, the ensuing debate about the ethics of surrogacy has, for the most part, remained within the contractarian paradigm of social relations. Some argue, with Judge Sorkow, in favour of surrogacy on the basis that contracts are just as binding for pregnant women as for anyone else; others seek to ban surrogacy altogether on the grounds that it leads to the exploitation of women. The Australian National Bioethics Consultative Committee (NBCC), in its 1990 report on surrogacy, takes the middle ground by recommending that surrogacy be permitted but that the surrogacy contract not be enforceable (NBCC 1990: 36). I have no argument with this conclusion. But it does depend on the same paradigm of social relations which led to the opposite conclusion in the case of Baby M. As a consequence, the NBCC saves the social contract by excluding pregnant women from it. This seems highly problematic. If the contractarian model of social relations, used widely in biomedical ethics, cannot fairly include pregnant women then the model itself requires closer scrutiny. What I will argue in the remainder of this chapter is that biomedical ethics in general and the surrogacy debate in particular, in forgetting the body and armed with its universal rules, relies on an inappropriate model of the relation between the individual and her body and misconceives the nature of the relation between the individual and others. In short, if ethics is to allow sexual difference, it is necessary to rethink embodiment and the nature of identity and difference.

SURROGACY AND THE FORGOTTEN BODY OF BIOMEDICAL ETHICS

I have said that biomedical ethics forgets embodiment in determining the ethics of practices such as surrogacy. However, with some work a concept of embodiment can be found within the paradigm of social relations used in these deliberations. The principles informing the NBCC's report on surrogacy provide an apt guide to what is typically taken to be the nature of the individual, and of relations between individuals, in biomedical ethics. These principles are spelt out as follows:

1 '[T]he *principle of personal autonomy* or self-determination, namely that people should have the right to make their own life decisions for themselves so long as those decisions do not involve harm to others';
2 'The *principle of justice*', namely that arrangements between individuals should not involve exploitation and should best serve the interests of all those involved (in this case those of the surrogate mother, the unborn child and the commissioning couple);
3 'The *principle of the common good,* namely that the good of the whole community must be considered' in arrangements made between individuals (NBCC 1990: 14).

There are two features about the individual which are assumed in, and allow the use of, these principles in determining the ethics of practices such as surrogacy: first that the individual is disembodied, and second that the individual's identity is given prior to its relations with others. In what follows I will draw out these features, which are only implied, and point to the difficulties they produce.

The NBCC is explicit about the kind of individual it assumes in its principle of autonomy and it is here we find a simplistic and problematic understanding of embodiment. Quoting from John Stuart Mill's essay *On Liberty*, the individual is defined by the dictum: 'over himself, over his own body and mind the individual is sovereign' (NBCC 1990: 15). In the only direct reference made to the body in this report, it is given the status of a passive object governed by an individual agent who somehow stands above it. But much more is implied about the body in this concept of the individual. Mill's concept, which can be traced back to the philosophy of John Locke, is one where the individual is said to have property in their own person (Locke 1967: 305–6). Locke had effectively severed the rational agent from his or her body giving the agent property rights over the body and the products of its labour. So, for the NBCC, autonomy implies the freedom to decide how to dispose of one's body so long as others are not harmed by that decision. By itself, this principle of autonomy would allow a woman to use her body for the purposes of a surrogacy arrangement.

What is also implied in this concept of the individual is that the individual agent is unified and present to her self: she has immediate access to her motives and desires, at least potentially, and can weigh up competing possibilities in arriving at a decision about what to do with this appendage called the body in order to best serve her interests. And, in the event that the individual is ignorant of her best interests, it is assumed that the ethicist can define these for her.

This notion of the disembodied individual forms the foundation for a certain understanding of the nature of relations between individuals introduced through the principles of justice and the common good. The NBCC's stated aim is to reconcile the principle of personal autonomy with those of justice and the common good (NBCC 1990: 23). And in this, the implicit focus of regulation are relations of contract and exchange between self-present individuals where the object of exchange is the individual's body. This model of exchange should not be surprising. Once the individual is said to have a property relation to her body, it is taken for granted that she has the right to exchange products of her body's labour, under contract, for financial or other reward.

The first point to note about the NBCC's paradigm of social relations is the atomism it implies. The individual's identity is given prior to her relations with others and prior to the rules which govern those relations. The individual is said to meet the other on equal terms and the contract they make constitutes their social relation. Secondly, what is assumed, when contracting out body property, is that the self does not, or should not, change (her mind) over the duration of the contract even though the terms of the contract may involve addition to, or subtraction from, her body. (In the case of Baby M, Mary Beth Whitehead was held to her contract on these grounds despite changing her mind about giving up her child.) The third point to note is that only some kinds of body property are allowed into the market place. What seems to be at issue in bioethics is the preservation of an assumed atomism and, implicitly, the significance of the body property being exchanged. So, while the principle of autonomy grants the right to freely contract out property in one's person, the principles of justice and the common good place ethical limits upon ways in which the body can become a legitimate object of exchange. A woman, like a man, has the right to do with her body what she will except, it would seem, if that body is involved in reproduction. The pregnant body is not a body which can be easily exchanged in the market place.

In the case of surrogacy, the most general object of exchange and regulation seems to be a woman's body. Those players said to compete for sovereignty over this body are: the woman herself (considered separate from her body), the unborn child, the commissioning couple, the bioethicist (as representative of the law and the common good), and the biomedical practitioner (if reproductive technology is used). The woman's autonomy (her right to sovereignty over her own body), is pitted against the possibility that her actions (and those of biomedical science) may bring harm to herself, to the child in the future, to others and to the fabric of society in general. The ethicist, using the principle of justice, may deem that the surrogate's autonomy is threatened by a biomedical practitioner (if reproductive technology is used) or by the commissioning couple. This perceived threat to the woman's autonomy is a common basis for feminist objections to surrogacy. Or the autonomy of others (the child, for example) may be deemed to be at risk in the future as an effect of this contract. On the basis of the principle of the common good, the surrogacy contract may appear to threaten the social fabric (if it is said to involve baby selling). So, in determining the ethics of a contract between individuals in general terms, the value and integrity of the individual's body, as assumed by the individual her- or himself, is weighed against the value and integrity of others and of the body of the community.

An immediate problem with this model of social exchange and its concept of the individual is that, as I've suggested, it doesn't seem to apply to pregnant bodies. On the one hand, if women are to be admitted into social exchange on the same basis as men, then we should have the right to participate in contracts to do with property in our person.

Yet, under the guise of being held to her contract, a woman can be forced to give up her child, as in the case with Baby M. This doesn't seem just to most observers, as evidenced by the outcry from feminists and others in the wake of this ruling. On the other hand, to exclude women from social exchange on the basis of our embodied womanhood would be contrary to the ideals of autonomy and feminism. As neither option is satisfactory, the contract model of social relations itself requires reviewing. To do this I will turn to some feminist critiques of this paradigm. The first is a general critique provided by Carole Pateman who challenges the atomised disembodied concept of the individual and the contract model of social relations. The second is a critique of the use of this paradigm in ethics from the work of Carol Gilligan and her followers.

FEMINISM AND CONTRACTARIAN ETHICS

In *The Sexual Contract* (1988) Carole Pateman takes issue with the contract model of social exchange both as a story about the origin of civil society and as a principle for relations within contemporary society. As Pateman explains, social contract theory claims that society is founded when insecure natural freedom is exchanged for equal (although restricted) civil freedom, a freedom and equality which is supposedly reproduced when entering into particular contracts. And, according to Pateman, that contract is given as the paradigm of free agreement is based on the atomised, disembodied concept of the individual (Pateman 1988: 55–7). That is, as the individual is said to be the proprietor of his person, then relations with others must be created in such a way as to protect this property right. Contracts supposedly allow for the use of another's property by mutual agreement, to mutual advantage and with security over time.

Aside from a general suspicion with the concept of the individual assumed in contract theory, Pateman takes issue with the contract model of social exchange on at least two counts. She argues that, contrary to the claim that contracts involving property in the person are to the mutual benefit of both parties, they actually constitute a relation of subordination (whether the contract is entered into voluntarily or not) (Pateman 1988: 55–9). For the purposes of her argument, Pateman points out that what is exchanged between parties to a social contract are words (which constitute a social relationship) and civil obedience for protection of property in a person. And, following the 'original' contract, the law provides security over time for contracts between individuals where the personal property of one party is rented or sold to another. The social relationships thus constituted involve subordination, according to Pateman, because the party who purchases or 'protects' the property held in another has the power to decide what the other must do to fulfil their side of the contract.

Pateman's second objection to the contractarian model lies in her claim that the individual agent assumed in it is male (Pateman 1988: 39–54, 59–60). Beginning with the distinction between a rational agent and their body, stories of the original social contract (with the possible exception of Hobbes) assume women do not have the rational capacities necessary for the social contract. They are parties to a sexual contract only, insofar as they are assumed to be property *per se*. This sexual contract (assumed to be natural rather than social) effectively gives men the right of access to women's bodies. And, Pateman suggests, while women have since been granted access to the same civil institutions as men, they become parties to social contracts only insofar as the property in their persons is devoid of anything specific to their womanhood. The sexed bodies of women and their products remain the property of men. Taking both criticisms together, Pateman concludes that the social contract effectively trades man's civil freedom for woman's social subordination.

These two criticisms inform Pateman's objections to the surrogacy contract (Pateman 1988: 209–18). There she argues that what is exchanged in the surrogacy contract is the genetic father's sperm and the surrogate mother's uterus. This appears to be a free and equal exchange because the masculine concept of the individual (as owner of property in one's person) has been extended to women, making sexual difference apparently irrelevant to reproduction: as each party's property is said to have the same value, the birth mother has no greater claim to the product of exchange. But, Pateman argues, the contract actually extends paternal rights and secures the subordination of women. This conclusion is based on her claim that in this contract only sperm has the status of actual property and, as its owner effectively pays for access to the surrogate's body (which is thus reduced to an empty vessel), he assumes ownership of the product of the contract.

While accepting, for the moment, Pateman's valuable critique of the sexual contract in general, I find her analysis of surrogacy problematic insofar as it depends upon the same notion of the individual she criticises elsewhere. Once the individual is understood to be a conscious agent separate from and owning property in his or her body and its products, then the decision about what part of that property is exchanged in a surrogacy contract remains fairly arbitrary, as does the subsequent decision about the fairness of that contract. Pateman claims that sperm and the uterus are the exchangeable commodities. But, as she acknowledges, contractarian defenders of surrogacy (the most notorious being in the case of Baby M) will just as easily argue that it is the services of the woman's body which are exchanged for financial reward without detriment to any person.

The NBCC settles for an entirely different division of personal property in its analysis of surrogacy, yet argues towards a similar conclusion to Pateman. For the NBCC, the child, rather than the mother's body, is the most pertinent exchangeable commodity (NBCC 1990: 29). Such a suggestion, in keeping with contractarian individualism and its model of exchange, takes the pregnant body to be two bodies. Surrogacy becomes a problem about competing claims between two individuals, the mother and the foetus, where both are assumed to be autonomous entities with the right to sovereignty over their respective bodies (the mother in reality, the foetus potentially). Again this division of the pregnant body is arbitrary and not necessarily legitimate. And even if it is, decisions about whose property rights should take priority become reduced to the question of whose side you are on. Pateman counters this kind of division of the body with the claim that, in the case of the pregnant body, the foetus is part of the self (Pateman 1988: 214–15). What I am suggesting is that the way that Pateman divides the woman's body for the purposes of her account is equally arbitrary and that there is something fundamentally problematic about dividing a self, pregnant or otherwise, into a conscious agent and a passive, divisible body. (I'll return to this point.)

Besides the assumption that we can divide the pregnant body into two autonomous entities, the more general atomism implied in the NBCC's notion of individual autonomy is also questionable. The NBCC expresses no doubts that the body is a unified entity separate from those who seek to govern its activity, implying that autonomous decisions are about one's body rather than being enmeshed within it. And it finds no problem with the distinction between personal autonomy (private morality) and the values constituting the common good (the legal sphere) (NBCC 1990: 15). This distinction implies that the individual comes before her social relationships, that what is exchanged in a contract (constituting those relations) is something apart from herself and that her decisions are uncontaminated by the values which constitute the 'common good'.

But the NBCC does entertain submissions which question this notion of autonomy. One submission, for example, opposes surrogacy *per se* on the basis that while women's autonomy should be upheld, free choice implies being fully informed of the consequences of a decision and yet, in the case of surrogacy or the use of reproductive technology, a woman can never be fully informed of her future emotional wellbeing (NBCC 1990: 17).[1] This implies that the distinction between a woman's agency and her embodied wellbeing over time is illegitimate. Another submission claims that women's disadvantaged social position influences and limits the choices they can make, leaving them open to exploitation by the commissioning couple, in the case of surrogacy, and by biomedical science, if reproductive technology is used (ibid.: 17). Here the suggestion is that individual decisions are already informed by the agent's social position (and hence by the values embodied in the common good). This argument against surrogacy insists, as feminist theorists do in general, that the patriarchal context be considered when evaluating the ethics of reproductive practices. But the argument is usually extended to a call to minimise those practices on the grounds that they increase patriarchal control over women's bodies.[2]

The importance of the feminist challenge to the distinction between the private sphere of autonomy and a (public) common good cannot be overstated. But too often this challenge leads to the disturbing conclusion that a woman's actions (said to be already limited in a patriarchal society) should be further constrained in the interests of her own protection. This conclusion implies that there is a protected space for women uncontaminated by patriarchal social values. And this can only be assumed by allowing back into the analysis the notion of individual autonomy as the exercise of sovereignty over one's body and the distinction between the individual agent and the social context. Not only does the conclusion to restrict women's actions contradict the terms of the analysis but it portrays women who participate in practices such as surrogacy as either passive, naive victims of patriarchy (being coerced or duped into fulfilling its needs) or active agents in the immoral reproduction of a system which exploits women. Further, as the line between 'true' autonomy for women and male control over women's bodies seems to be drawn at reproduction and sexual relations with men, then this argument risks making heterosexuality and pregnancy unethical *per se*. And, as the line is drawn such that contractarian concepts of the individual and of relations between individuals remain ultimately intact, then the social-contract model of social relations is preserved by excluding women, insofar as our bodies signify womanhood.

This is also the effect of the NBCC's assessment of the surrogacy contract. A space for women's autonomy is apparently preserved by deeming surrogacy a legitimate expression of a woman's right to sovereignty over her own body. The 'common good' (said to be distinct from private agency) is also preserved from the perceived threat of baby selling by restricting the profit which can be extracted from the contract. And the woman's future wellbeing is also preserved by removing any legal weight from the surrogacy contract. While the validity of the surrogacy contract is suspended, the social contract as a paradigm of free and equal exchange is upheld, based as it is on an arbitrary division of the body, on a problematic distinction between the agent and her body and by a questionable distinction between the spheres of personal autonomy and the common good. Yet it is only upheld by making pregnant women an exception to the rule. It would seem that in pregnancy the distinctions don't apply: a pregnant woman is autonomous if she gives nothing of herself away, she is deemed more susceptible than other women to exploitation in the face of the common good and is more prone to change her mind as her body changes. But if the distinctions which uphold the social contract are not applicable to the pregnant body then perhaps this is also the case with any body.

Pateman, with justification, wants to rid ethico-political analysis of contract as the paradigm of social relations. But, without a viable alternative, her own account of the subordination of women within the contract schema tends to leave women out of social exchange. Within her argument against the surrogacy contract, for example, she goes so far as to question the ethics of non-commercial surrogacy (which presumably would involve no formal contract) (Pateman 1988: 211). She seems to want to reject surrogacy understood as a gift (ibid.: 211). Is Pateman justified in moving from the argument that contracts imply the subordination of women to the conclusion that women should not participate in activities normally thought of in terms of contract? I think not. For the reasons discussed above, it seems inappropriate to restrict the activities of women on the basis of anything specific about women's embodiment without offering a different way to think the nature of social exchange. I will now turn to explore one alternative model provided by feminist ethics.

FROM CONTRACT TO CARE

The surrogacy debate for the most part ignores a well-established philosophical and feminist tradition which is critical of a contractarian model of social relations. The main point at issue is Locke's concept of the individual and of relations between individuals. Pateman, as I've suggested, blames the spurious notion of free and equal contracts on Locke's idea that the individual owns property in his body (property which requires protection and which can be sold under contract). Not only does the idea of contract assume a sharp and problematic distinction between the individual agent and their body but it also assumes that the agent should or does remain the same over time even if the body changes. This idea can also be attributed to Locke who located personal identity in the continuity of the same consciousness throughout minor or dramatic corporeal changes. But underlying both these claims about the relation between the agent and their body is a foundational and problematic understanding of identity and difference. According to Locke, an entity is identical with itself if it has the same origin in time and space 'and that which had a different beginning in time and place from that is not the same but divers' (Locke 1975: 328). This principle of individuation not only underlies Locke's claim that self-consciousness (rather than different moments of consciousness or the body) allows for personal identity over time but it also makes contract the basis of social relations. The claim is that the individual already has a self-contained identity (based on the continuity of consciousness) prior to their relations with others. A contract with another is said to establish a social relationship without affecting the respective identities of the different parties involved. From this understanding of identity and difference it can be claimed that a surrogacy contract involves exchange of a service for money or the mutual exchange of bits of bodies without affecting the identity of the parties to the contract. Or from the same model of identity and difference opponents to the surrogacy contract imply that a woman's autonomous social identity is preserved if she keeps her body to herself.

It is this concept of identity and difference which has been subjected to thorough criticism at least since Hegel. Hegel had claimed that identity is a product of, and carries within it, its relation to what is different and that no existence of any kind conforms to the (Lockean) maxim that everything is identical with itself and that difference is an external relation (Hegel 1975: 166–7). I will be discussing Hegel's formulation of identity and difference in more detail in chapter 3. It suffices to say for the moment that Hegel's challenge to self-present identity implies a different assessment of the form and function of exchange in social life. What is suggested is that identity is always relational, always a product of exchange rather than prior to it. It is this kind of idea of the interrelatedness of self-identity which informs Carol Gilligan's ethics of care. I will now consider her approach to ethics, aspects of its effect upon feminist ethics and whether care or its derivatives overcomes the problems I have raised.

In her now familiar work, *In a Different Voice* (1982), Carol Gilligan issues a challenge to moral theories which privilege justice and rights, theories where maturity is measured in terms of the ability to solve moral dilemmas using universal ethical principles derived from the contract model of social exchange. Lawrence Kohlberg (1971) had developed such a theory from his empirical research into cognitive moral development. In his schema of cognitive capacities he ranks the ability to abstract oneself from one's context to impartially apply principles of justice (such as the right to equal consideration) above the ability to empathise with others within particular interpersonal relations. Gilligan, with Murphy (1980), reformulates this distinction in terms of formalism (ethical orientation of justice and rights) and contextualism (ethical orientation of care and responsibility) and challenges the hierarchy Kohlberg had established between them. What is of concern to

Gilligan is that women seem to score badly using Kohlberg's system of moral development (which, significantly, is based on the behaviour of boys) (Gilligan 1982). According to the results of Gilligan's studies on both sexes, women's moral judgement tends to be contextual rather than formal. When making moral judgements women tend to combine 'the recognition of interconnection between self and other with an awareness of the self as the arbiter of choice' (Gilligan 1982: 96). That is, women tend to reject the concept of individual rights in favour of responsibility and care (ibid.: 105). In Kohlberg's schema, this tendency would be viewed as a sign of immaturity. But in Gilligan's revision of stages of moral development the ability to consider context, the details of relationships and the viewpoint of the particular other is seen not as moral failure but as essential to moral maturity.

I will not dwell on the details of the debate between Gilligan and Kohlberg, nor on the complexities of their various positions. This has been well documented elsewhere.[3] What I want to highlight is that Gilligan's critique exposes a certain relationship between moral theory and models of identity and difference: giving priority to moral reasoning based on rights and justice assumes atomised individualism whereas giving priority to responsibility and care assumes an essential interconnection between self and other. Further, the assumption of isolated self-presence allows for an ethics based on universal principles whereas the assumption that identity is constituted through a social relation to what is different allows for consideration of social context and the particularities of interpersonal relations. Without rejecting universalist moral theories based on atomistic individualism, Gilligan at least wants to give equal validity to an ethics of care based on the recognition of a fundamental connection between self and other.

What is unclear is whether Gilligan's ethics of care alone can accommodate sexual difference within ethics. Her work is psychologistic and descriptive – it describes differences in modes of moral reasoning between some men and some women. But are these differences natural or socially constituted? Can they be generalised? And, if so, can care be legitimated with the stroke of a pen? Acknowledging the work of Nancy Chodorow (1978) on the social constitution of sexual difference, Gilligan gives weight to the possibility that the differences between the sexes she observes in her study reflect the process whereby separation from the mother and individuation are stressed in the development of boys whereas attachment and intimacy are encouraged in the development of female gender identity (Gilligan 1982: 8). But for the most part Gilligan eschews a critical analysis of the genesis of the differences she observes between the sexes. And in valorising an attitude of care in women as a complement to a masculine ethics of justice, Gilligan risks entrenching those familiar gender stereotypes which some feminists have blamed for the continuing oppression of women. Again, as these problems have been widely discussed, I will take them no further here.[4] Rather, I will now turn to some recent attempts to critically appropriate Gilligan's work.

The point at issue is that if the atomised, disembodied individualism which underlies the contract paradigm of social relations is inappropriate for dealing with the ethics of pregnancy or anything specific to women's embodied experiences, then will a paradigm of social exchange based on the interdependency of self and other be better able to cope? Seyla Benhabib, in her astute discussion of ethics and gender in *Situating the Self* (1992), appears to think so. At least, she attempts to establish a way to deal with gender difference in general terms by taking account of the interactive nature of self-other relations. In this she follows Gilligan, with qualification.

Benhabib, like Pateman, is critical of contractarian theories of social relations. She argues, in particular, that the ideal of autonomy in Kohlberg's moral theory and in universalistic contractarian theories leads to a 'privatization of women's experience and to the exclusion of its consideration from a moral point of view' (Benhabib 1992: 152). She attributes this exclusion to the disembodied and disembedded concept of the moral self assumed in these theories (by 'disembedded' I take her to mean atomised or detached from others). The self of social-contract theory is, as Benhabib explains, a man, alone in the state of nature and even independent of his mother (ibid.: 156). In the public sphere of justice, the self is also a man (white and middle class); woman is confined to the private sphere and defined as what man is not (ibid.: 157). When this male individual enters culture and meets his other for the purposes of social exchange and consensus, this other is viewed as a narcissistic image of the self. This conception of the other, which predominates in universalist moral theory, is the 'generalized' other, as Benhabib puts it. We assume that what constitutes the moral dignity of the other 'is not what differentiates us from each other, but rather, what we, as speaking and acting rational agents, have in common. Our relation to the other is governed by the norms of *formal equality* and *reciprocity*' (ibid.: 159). But, as Benhabib has suggested, as the self is male (white and middle class) and the other is a mirror image of the self, then the paradigm of the 'generalized other' excludes consideration of differences arising from anything specific about women's experiences.

In contrast to the generalised other, Benhabib posits the 'concrete' other, the other understood as 'an individual with a concrete history, identity and affective-emotional constitution' (Benhabib 1992: 159). The identity of the 'concrete other' is distinguished from the self on the basis of sex, race, class, cultural differentials and abilities (ibid.: 164). In our relation to the concrete other we seek to comprehend the other's specific desires; our 'differences in this case complement rather than exclude one another' and the norms of our interaction are friendship, love and care (ibid.: 159). The ability to consider the concrete other is the ability which Gilligan had attributed to women.

Benhabib provides a convincing argument for her claim that universalist moral theories, by privileging an autonomous self and 'generalized other' at the expense of a 'concrete other', suffer from internal inconsistency and incoherence. In other words, if universalist moral theories ignore the concrete other, we are left without a means of judging whether our moral situation is the same as or different to that of the other (Benhabib 1992: 164). Universalist moral theories which assume a 'generalized other' are not universalist at all.

Benhabib does not simply valorise the standpoint of the 'concrete other' nor women's said capacity for care. What she seeks, more explicitly than Gilligan, is a universalist moral theory based on an integrated model of ourselves and others as 'generalized' as well as 'concrete'. It is this move I find problematic. But it is a view shared by others in the wake of Gilligan's work. Elizabeth Porter, for example, in *Women and Moral Identity* (1991), a comprehensive study of gender and moral theory, advocates a 'philosophy of synthesis' which would transform our dualistic world into a dialectical one; a synthesis or reciprocity between moral principles and contextual adaptability; a recognition of many voices (Porter 1991: 49, 165, 169). While Porter and Benhabib differ in a number of ways, the general issue their work raises is the possibility of integrating (formal) reasoning based on moral principles and the 'generalized other' with consideration of context and the differences issuing from the 'concrete other'. Both argue for this possibility without insisting that women are better at the latter.

Benhabib provides a convincing argument (contra Habermas and Kohlberg and following Gilligan) for the inclusion (along with issues of justice) of the spheres of kinship, love, friendship and sex within the moral domain. Widening the moral domain to include issues of care (validated from an impartial perspective), respecting the moral dignity of the 'concrete other' and accounting for differences through individual deliberation and open collective discussion will, according to Benhabib, generate a genuinely universalist ethics (Benhabib 1992: 179–90). While I accept her argument for extending the moral domain to include care for and consideration of the 'concrete other', what is less convincing is that this 'interactive universalism' will account for differences as Benhabib hopes. There is little in this proposal to integrate justice and care, formal principles and context, which provides a safeguard against the possibility that the differences of the 'concrete other' are subsumed within those of the 'generalized other' under the guise of 'collective dialogue'.

There are two omissions from Benhabib's discussion which leave me doubting whether 'interactive universalism' overcomes the exclusions generated in the name of the 'generalized other': there is no substantial account of how the embodied identity and difference of the other is generated; and, connected to this, there is no account of how my 'dialogue' with concrete others may contribute to, transform and efface their differences.

On the question of the constitution of embodied identity, Benhabib does acknowledge that sexual difference is socially constituted and embodied: 'the self develops an *embodied* identity, a certain mode of being in one's body and of living that body' through a 'gender-sex' system, 'the social-historical, symbolic constitution, and interpretation of the anatomical differences of the sexes' (Benhabib 1992: 152). But, if it is the case that one's identity is embodied, sexed and an effect of one's specific socio-historical context, and I believe it is (in ways I will develop in later chapters), how can one abstract oneself from this specific embodied being to include consideration of the differences of the 'concrete other'? It turns out that, while Benhabib grants that the self is embodied, this is only significant in terms of our vulnerability and dependency as bodily selves (ibid.: 189). This takes no account of how the social significance of the body becomes the ground for exclusion from social exchange.

Benhabib's acknowledegment that identity is sexed, embodied and specific to one's position within a differentiated social text is subsumed under her dominant thesis that the identity of the self is constituted by/as a 'narrative unity' through interaction with a community of others (Benhabib 1992: 5–6). On the one hand it is this interaction, constitutive of the self, which implies that self-identity is always dispersed into others. On the other hand, Benhabib (following Habermas and Arendt, with qualification) makes interaction the basis for the possibility of a 'moral conversation' where perspectives are reversed, the other's point of view is taken, different voices are accommodated and agreement is reached (ibid.: 8–9, 76–82, 104–5). For the purposes of this proposal, interaction seems to take place between self-present, disembodied minds: 'interactive rationality' and speech are the instruments for a 'moral conversation' which requires that the parties be transparent to themselves and to others (ibid.: 129). So, despite disclaimers to the contrary, Benhabib's interactive universalism does seem to rely on forgetting the body and the possibility (which I will argue for in chapters 5 and 6) that perspectives are embodied.

Significantly, Benhabib does not test her thesis with a concrete case involving anything specific about women's embodiment (such as pregnant embodiment). Rather she chooses examples of conflicting economic and career interests among friends and siblings (Benhabib 1992: 128,

184–6). As my discussion of surrogacy indicates, it is more difficult to accommodate activities involving sexed bodies within a paradigm which assumes that the self and the other are housed in male bodies, than it is to resolve conflicts where the interests of the parties do not obviously relate to their embodiment.

On the question of the nature of the self-other relation and the status of the other's difference, Benhabib insists that the self is not an atomised individual, with an identity given separate from others. If identity is constituted in relation to others, then how can the self be transparent to itself and to others? And does not one's interaction with the other affect and transform their identity and difference? While I will develop these points more fully in chapters 3 and 4, I raise them now to mark the possibility that, despite her disclaimers, Benhabib's 'moral conversation' is really a monologue where agreement is reached when the point of view of the 'concrete other' is subsumed under that of the 'generalized other'.

The importance of Benhabib's insistence that the 'concrete other's' point of view be accommodated in ethics through common action and debate cannot be overstated. That 'interactive rationality' will achieve this end is what I question. To illustrate why I don't think 'interactive universalism', a philosophy of synthesis or the integration of formal principles and context would get us far in putting sexual difference into ethics, I will return to the problem of the surrogacy contract. As neither Benhabib nor Porter deal with the ethics of reproduction, I'll refer to the work of Susan Sherwin which does.

Sherwin (1989), like Benhabib and Porter, follows Gilligan's lead in emphasising the importance of considering context, interpersonal relations and caring in ethics. She rejects the isolated individualism underlying contractarian approaches to ethics and recognises that the person is a product of their specific social context (Sherwin 1989: 64). While she sees the need to accommodate differences arising from contextual specificity, she, like the other theorists discussed, opts for 'maintaining a certain level of generality in ethics' (ibid.: 65). Accounting for both context and generalities is Sherwin's mode of 'interactive universalism'.

Sherwin's concern is with normative medical ethics in general and the ethics of reproductive practices in particular. She suggests that if we take a context-specific approach to practices such as surrogacy, abortion and the use of *in vitro* fertilisation, we can appreciate that the choices women make to participate in these practices are embedded in their lives and interpersonal relations (Sherwin 1989: 65, 67). We can sympathise with the woman's specific circumstances and desires and give moral weight to the choices arising from these circumstances. But, Sherwin argues, this attention to context must be mediated by appreciation of the wider context, the general social pattern constituted by these practices. The generality which must be considered, according to Sherwin, is the tendency of the male-dominated medical profession to control women's personal and reproductive lives (ibid.: 66). While this claim is not an explicit formal moral principle it does imply the principle of freedom from interference or the right to sovereignty over one's body. With the introduction of this generality, practices which increase patriarchal control over women's bodies are discouraged (surrogacy, reproductive technology, and 'coerc(ing) women into seeing an unwanted pregnancy through' (ibid.: 67)) while those which don't (abortion if freely chosen) are endorsed.

Sherwin's is a seductive argument for those who seek to validate abortion but disqualify surrogacy and the use of reproductive technology. But, as with Benhabib, I am concerned about the disappearance of context or the 'concrete other' once a formal principle is evoked. Further, Sherwin's generality reinstitutes assumptions about the individual and her social relations which were discounted in her critique of contractarian ethics. That is, in attempting to minimise medico-patriarchal control over women's reproductive lives, she assumes there exist choices about reproduction which are freely made and uncontaminated by patriarchal medical discourse; she assumes atomised individualism. Connected to this, and perhaps more disturbing, is that in claiming that specific instances of reproductive practices constitute a general pattern of attitudes which undermine the social position of women (Sherwin 1989: 65), she risks blaming women who participate in these practices for the social subordination of women in general. And, as discussed above, it is not clear in this kind of argument whether any sort of reproductive practice could be considered free from patriarchal control and therefore ethical. In other words, Sherwin risks reproducing the paradigm she eschews by excluding pregnant embodiment from valid social exchange.

I do not want to deny the importance of the work of these theorists in developing an ethics of care: they have clearly revolutionised ethics in their attempts to include consideration of differences. What bothers me is the tendency for the assumptions of self-present disembodied individualism to creep back into their theses despite various disclaimers and with the risk of effacing the differences they seek to accommodate. This tendency arises I think from paying insufficient attention to the implications of two important claims, claims which are made by the theorists mentioned but which remain undeveloped in their work. The first is that the self develops an embodied (and sexed) identity within and as an effect of a specific social context (Pateman 1988: 16–17; Benhabib 1992: 152; Porter 1991: 16, 21). The second claim is that one's identity is constituted in the context of and through relations with others (Benhabib 1992: 5; Porter 1991: 22; Sherwin 1989: 69).[5] What is meant by both these claims determines the field of an ethics of difference. Therefore these are the theses to which I will give further attention.

In attempting to answer the question of what is meant by embodied identity and what is the nature of identity and difference, my aim is to develop the means of dealing with different modes of embodiment (such as pregnancy) and embodied practices (such as surrogacy) without excluding women from social exchange. The understanding of identity and difference I will develop in chapters 3 and 4 suggests a model of social exchange based not on contract but on 'the gift' of part of oneself to another. Within this paradigm of social relations surrogacy is not unethical but profound. The concept of embodiment I will raise in chapter 2 and develop in chapters 5 and 6 allows for that which contract theory eschews: the possibility that dramatic changes in embodiment effect changes in the self over time. This invalidates a surrogacy contract, for example, without either banning surrogacy or excluding women from exchange in general.

But I am running ahead of myself. To entertain a rethinking of embodiment and the constitution of identity and difference under the umbrella of ethics, alters what is meant by ethics. In the following chapter I will justify this shift in focus from a study of moral principles and moral judgement to a critical study of that which constitutes our embodied place in the world.

Chapter 2

Ethics, embodiment and sexual difference

I suggested at the close of the previous chapter that, in order to put sexual difference into ethics, it is necessary to account for (a) the social constitution and significance of embodiment (sexed embodiment in particular) and (b) the genesis of identity and difference. What follows is both an elaboration of what I mean by the social constitution of embodiment, with reference to the work of Michel Foucault, and a justification for the inclusion of this problematic under the umbrella of ethics. In the second section of the chapter I will discuss the need to rethink the production of individual identity and difference.

A GENEALOGY OF ETHICS

We usually think of ethics as either the study of the logical status of our moral judgements or as setting down a set of universal principles for regulating behaviour. In focusing on moral principles and moral judgement the assumption is that individuals are present as self-transparent, isolated, rational minds and that embodied differences between individuals are inconsequential. However, as I have argued in the previous chapter, this understanding of ethics and its assumptions about the individual disqualifies women from ethical social exchange insofar as our bodies signify womanhood.

Even if we grant that ethics is about moral principles and moral judgement, it is also about location, position and place. It is about being positioned by, and taking a position in relation to, others. Being positioned and locating others requires embodiment and some assumption about the nature of the place from which one moves towards others. It should not be surprising then that 'ethics' is derived from the Greek word *ethos*, meaning character and dwelling, or habitat.[1] Dwelling is both a noun (the place to which one returns) and a verb (the practice of dwelling); my dwelling is both my habitat and my habitual way of life. My habitual way of life, ethos or set of habits determines my character (my specificity or what is properly my own).[2] These habits are not given: they are constituted through the repetition of bodily acts the character of which are governed by the habitat I occupy. From this understanding of ethos, *ethics can be defined as the study and practice of that which constitutes one's habitat,* or as the problematic of the constitution of one's embodied place in the world.

The discrepancy between this approach to ethics and that based on universal principles is not simply a question of etymology. Related to this are different, and usually unacknowledged, understandings of the components which go to make up our spatio-temporal being-in-the-world. The difference pertains to whether we think our 'being' is composed primarily of mind or matter; to what we understand by the relation between mind and matter; and to whether we think the world we inhabit is homogeneous or fragmented. Underlying all these questions is some assumption about the meaning of 'in'.[3] An ethics based on universal rational principles assumes that our 'being' is a discrete entity separate from the 'world' such that we are 'in' the world after the advent of both. An ethics based on the problematic of place, on the other hand, claims that our 'being' and the 'world' are constituted by the relation 'in'. In other words, the understanding of ethics I am evoking recognises a constitutive relation between one's world (habitat) and one's embodied character (ethos).

I have also suggested that, besides an understanding of the constitution of embodiment, it is also necessary to consider the effect our relation to another may have upon the constitution of our ethos (and vice versa). This is necessary because to belong to and project out from an ethos is to take up a position in relation to others. This involves comparison, relation to what is different and to what passes before us. Taking up a position, presenting oneself, therefore requires a non-thematic awareness of temporality and location. And the intrinsic reference point for temporality, spatial orientation and therefore difference is one's own body. Taking a position in relation to others again involves some reference to embodiment, the significance and specificity of which comes together with ethics by virtue of our spatio-temporal being-in-the-world. But if ethics is about taking a position in relation to others then it is also about the constitution of identity and difference.

Despite the dominance of an ethics which emphasises rules of engagement for relations between self-present individuals, there are some contemporary accounts of an ethics based on the problematic of the constitution of one's ethos.[4] These variously locate the body as the locus of one's ethos – where the body is constituted by a dynamic relation with other bodies in a social context of power, desire and knowledge. Such an approach to ethics takes into account how the individual is constituted within a social context and allows for the possibility that differences may arise. Perhaps the best known of these accounts is that provided by Michel Foucault. In turning to Foucault's work I am highlighting one of the two problematics raised in my reformulation of ethics: the constitutive relation between embodied place and the social world. (I will consider the need to account for the constitution of identity and difference in the second section of this chapter.)

Before discussing Foucault's important contribution to ethics, it is worth reiterating a problem I raised in chapter 1. There I discussed how universalist contractarian ethics cannot fairly accommodate sexual difference. In assuming a self-present mind governing a (white, middle-class) male body, contractarian ethics cannot deal with the moral status of sexed modes of embodiment: pregnant embodiment and the ethics of reproductive practices are left out of account. Feminist critics of moral theory partly attribute this exclusion of women's experience in general, and of reproduction in particular, to the (incorrect) assumption that the self is isolated from the social world and carries common attributes and interests. These critics call for the consideration of how the patriarchal social context constitutes embodied and sexed identities and how patriarchal (and medical) discourses perpetuate the subordination of women. Yet, as I have suggested, some of these same critics seem to assume (in attempting to secure women's freedom by, say, rejecting contracts to do with reproduction) that there is an outside to the patriarchal social context. Part of my purpose in elaborating Foucault's approach to ethics is to show that it is problematic (at least without some detailed discussion/explanation) to hold that on the one hand sexed identity is embodied and socially constituted, but that on the other hand one can either abstract from this embodied identity to take the other's position or secure freedom by moving outside the social context of which one is an effect.

In his later work Foucault turns to ethics. His aim, consistent with that of the theorists discussed so far, is to open up ethics to the consideration of different modes of being. This requires, he thinks, an aesthetics of self which in turn rests on a certain understanding of how one's embodied ethos is constituted. I find Foucault's elaboration of the social constitution of one's embodied ethos invaluable (although not without its problems). In *The Use of Pleasure* he makes a distinction between morality, as 'a set of values and rules of action', and ethics (Foucault 1987: 25–6). He defines 'ethos' as a manner of being and 'ethics' as a practice, a technique of self-formation (Foucault 1984: 377). Rather than establishing a set of rules of action Foucault attends to the ethics of the techniques by which one's ethos is constituted. One's manner of being or ethos has four components: an ethical substance, a relation to the moral code, some kind of self-forming activity and a goal of this activity (Foucault 1987: 25–32). Through a comparison of Ancient Greek and Christian ethics, Foucault proposes that the ethical substance is corporeal and that techniques of self are either normalising or aesthetic, depending on one's relation to the disciplinary moral code, and either oppressive or not, depending on one's relation to the other.

In evoking the possibility of aesthetic or creative techniques of self, Foucault, in his ethical turn, attempts to address his earlier concern with contemporary modes of subjection – modes of subjection which devalue and efface differences through normalising techniques of self-formation. One's relation to a disciplinary moral code, while not formulated as constitutive of an ethos there, is the mode of subjection analysed in *Discipline and Punish* (1979). Our dominant mode of self-forming activity is the topic of *The History of Sexuality Volume One* (1978). I will first of all elaborate Foucault's model of the constitutive relation between one's ethical (corporeal) substance and the moral code (or social context), and the techniques of self-formation which mediate this relation, before addressing his proposal for an aesthetics of self.

In his early studies Foucault located embodiment as the site of one's ethos and attempted to answer the following question: 'What mode of investment of the body is necessary and adequate for the functioning of a capitalist society such as ours? . . . One needs to study what kind of body the current society needs' (Foucault 1980: 58). His suggestion is that bodies are made, not given, and that they are made to fit properly within a certain social structure. By attending to the relation between embodiment and the disciplinary moral code, Foucault is suggesting that the moral code does not simply govern relations between self-present individuals but constitutes the embodied ethos of those individuals. And his thesis is that for the most part we are subjected, dominated and controlled via the very social discourses and practices which purport to guarantee our freedom.

What Foucault finds in his own study of what kind of body our society needs is the operation of micro-techniques of power which constitute individuals within a political and economic field of domination and control. The most effective mode of power he finds operating in the modern bureaucratic state is not a repressive exercise of power originating in the state, individuals or classes – although each of these exercises power by structuring and limiting the field of action of others.[5] There is another kind of power operating, based on a political rationality which promotes the health and welfare of the individual to ensure the welfare of the whole.[6] This rationale evokes a moral code and distributes techniques of power through knowledges (from the natural and social sciences to popular culture) and certain social practices (institutional and non-institutional). These do not operate to prohibit the activities of individuals said to be autonomous. Nor do they work on minds to produce socially acceptable beliefs. Rather, these techniques of power are productive – they operate on the body to transform it, divide it, invest it with capacities and train it to perform certain functions (Foucault 1979: 25). It is the body which is the locus of self-formation: an individual ethos is constituted via work on the body. The individual subject is produced through this operation of power in two senses: he or she is subjected to the actions of others (where the body is the object of disciplinary power) and he or she attains a social identity through self-knowledge (where the body is the object of self-reflection and the subject of confession) (Foucault 1982: 212). While these two aspects of self-formation can only be separated for heuristic purposes, Foucault tends to focus on disciplinary power (which effects a normalising relation to the moral code) in *Discipline and Punish*. He attends to self-decipherment, as a self-forming activity, in *The History of Sexuality Volume One*.

The constitution of the publicly viable body fits well with Foucault's analysis of the objectifying practices of surveillance and discipline in *Discipline and Punish*. There he argues that discipline and surveillance (which proceed through supervision, the collection of data and a system of penalties) are techniques of power which construct individuals who are compatible with our social and economic system. Individual bodies are *objects* of this operation of power in that they become docile and productive under its exercise. And they are the *instruments* of power in that they become complicit in the discipline and surveillance of both themselves and others (Foucault 1979: 170).[7] As an object and instrument of this productive power, men in public spaces, for example, need no coercion: their ethos becomes co-extensive with the body politic and they take their roles to heart. If the common interest becomes embodied in an individual ethos in this way, the ability to abstract oneself from this ethos if not impossible cannot be assumed.

The reproductive practices discussed in chapter 1 do not escape this web of discipline and surveillance. Both biomedicine and bioethics play pivotal roles in this constitution and normalisation of embodied habitats. Foucault claims that 'the health and physical well-being of the population' emerged in the eighteenth and nineteenth centuries 'as one of the essential objectives of political power' (Foucault 1980: 170). The nexus between health and political power arose through concern over the maintenance of a productive labour force. As a consequence, biomedical science has emerged entangled with other knowledges and practices which effectively divide bodies around a norm. The healthy are divided from the unhealthy in terms of what is useful and what is 'less amenable to profitable investment' (ibid.: 172) The health of bodies has become a question of economic management: without any obvious subject of power and with the co-operation of all of us, our embodied ethos is constituted via these kinds of practices to be compatible with the wider habitat we occupy.

Foucault's objection to this political investment of bodies and different parts of a body is not just because of its normalising and controlling effects. He also objects to the unequal distribution of forces it produces. According to his analysis, individualisation in institutions proceeds through the segmentation of time and space (Foucault 1979: 142–62). The time and space segments occupied by individuals are codified and ranked; each has a particular meaning and value and will incur different degrees of discipline and observation. An individual is identified and classified, and takes on this ethos, according to the code and rank of the space occupied. Discipline and surveillance as techniques of self-formation do generate different habitats but these differences are ranked according to prevailing norms. So our social practices not only produce docile, self-regulating bodies but integrate these into a hierarchy of domination, control and mutual dependence. The pregnant woman, for example, is subjected to more public scrutiny than the medical practitioner who directly monitors her. The power relation is asymmetrical. Yet the practitioner is not an intentional agent of power and their client is not (usually) the object of exploitation, prohibition or coercion. The relation may be asymmetrical and caught within processes of social control, but as it is in keeping with the rationale of communal welfare neither party can easily extract themselves from it without attracting the condemnation of the community.

In paying attention to the constitutive relation between a disciplinary moral code and one's embodied ethos, Foucault highlights modes of subjection which do not require prohibitive laws or coercion. It is this operation of productive power which is missed by an ethics intent on seeking out, and protecting our said autonomy from, repressive operations of power. In assuming that there is a space to exercise the sovereignty we are said to have over our bodies, outside of and against repressive power relations, the ethicist inadvertently allows the work of disciplinary power to operate more effectively. Further, a universalist ethics, by purporting to uphold the values of the 'common good', by policing norms embedded in a rationale of communal welfare, is also complicit in our subjection.

While we don't take our subjection lying down, Foucault's thesis on the constitution of our embodied ethos does point to the difficulties encountered in redressing inequalities. As power is productive as well as prohibitive, on his model, social practices which are assumed to be egalitarian (or non-prohibitive) reproduce inequalities at a material level. And as these discourses and practices transform bodies as they inform them with meaning and value, it cannot be assumed that the norms involved in the ranking of differences can be opposed by a liberating discourse of Truth. The truth of the self does not lie outside the ethos constituted by productive power. So, contrary to an assumption underlying some critiques of contractarian ethics, there would seem to be no high moral ground where the individual can exercise agency outside of the social codes which constitute desires asymmetrically. In fact, according to Foucault, it is the assumption that domination is secured by prohibition and that truth is external to power which has encouraged a second technique of self-formation which works together with disciplinary power: what Foucault calls a 'hermeneutics of desire' or the technologies of self-decipherment and confession.

Foucault discusses these self-forming activities in *The History of Sexuality Volume One*. According to his analysis, we have come to believe not only in an original autonomous space for ourselves but also in a hidden meaning in ourselves and others. We believe that, by knowing and asserting the truth about ourselves, we can resist domination by the state, by institutions and at an interpersonal level. But introspection and self-affirmation (about one's health, desires, crimes and sexuality) are not liberating. Confession is a 'truth effect' of power. It presupposes interpretation by others and serves to further mark us out for surveillance and self-regulation. For, as Foucault suggests:

> the agency of domination does not reside in the one who speaks (for it is he who is constrained) but in the one who listens and says nothing; not in the one who knows, but in the one who questions and is not supposed to know. And this discourse of truth finally takes effect, not in the one who receives it, but in the one from whom it is wrested.
>
> (Foucault 1978: 62)

Foucault is claiming that there is a disjunction between self-knowledge and interpretation of that knowledge by another; that truth will be decided by the interpreter, not the speaker; and that there is a material effect operating in confession such that your embodied ethos becomes what the interpreter says you say you are.

This analysis of confession, as a normalising mode of self-formation, highlights our complicity in our own subjection in the face of institutional agents of power (doctors, psychiatrists, teachers etc.). But it also points to how the assumptions of autonomous agency, self-transparency and transparency of the other contribute to the normative constitution of an ethos rather than liberating us from it. Self-decipherment and confession do not follow on after and reveal the truth of the one who speaks; that truth is constituted in the body of the spoken word.

The dissonance between a discourse of self-reflection and that of interpretation by a listener, and the constitutive effects of both, also presents a problem for an ethics based on the kind of 'interactive rationality' discussed in chapter 1. In claiming the ability to account for the other's difference through dialogue, such an ethics assumes that there is a core to the self and to the other which is both transparent to the self and remains unaffected by the dialogue. Or, it assumes that both parties begin their communication from equal positions such that neither is an agency of domination. Yet, if this is the case, if there is no difference in the meaning and rank of the ethos of the two parties, then such a dialogue is unnecessary. Or, if there is a difference, then the constitutive effects of dialogue are such that the difference would be subsumed under the norms within which the dialogue takes place.

While Foucault's account of the social empowering of bodies suggests a need to rethink the ethics of subjection, it is not immediately clear that it can account for sexual difference. He would no doubt agree that the sexed body is not a biological fact. Sex (understood as bodily pleasures and the acts associated with them) is certainly not a natural fact. For Foucault, pleasures and desires are incited not given; sex is a body unity constituted by discourses on sexuality (Foucault 1978: 154–7). Sex is the naturalised product of a moral code which, through techniques of discipline, surveillance, self-knowledge and confession, organises social control by stimulation rather than repression (Foucault 1980: 57). Similarly, as a number of feminist scholars have suggested, the sexed body (understood as that which carries the marks of sexual difference) is constituted and encoded with meaning and value according to prevailing norms and discourses.[8] Or, as Judith Butler suggests, partly following Foucault, gender is that embodied identity constituted through a 'stylised repetition of acts' the significance of which is social rather than natural (Butler 1990: 140).

Given that women have been traditionally assigned different habitats to men, the style and significance of our embodiment necessarily differs, as does the kind and amount of surveillance our bodies attract. If we grant, for example, that pregnant women (and I confine my discussion to white, western, middle-class) have been traditionally confined to the private sphere then we can assume, if we follow Foucault, that the maternal body carries a different social significance and a different normalisation process to the publicly viable body. The significance of the maternal body differs from the public body in that it is *the site of the reproduction of the social body*. And if we grant that inhabitation, discipline and normalisation of other bodies destined for public spaces begins at birth, if not before, then it should not be surprising if there is a large investment in monitoring reproductive and child-care practices. As Foucault has attempted to show, in the medical politics which arose in the eighteenth century, the family became 'the first and most important instance for the medicalisation of individuals. The family is assigned a linking role between general objectives regarding the good health of the social body and individuals' desire or need for care' (Foucault 1980: 174). Perhaps it is because the pregnant body is the site of the reproduction of the 'normal' body that pregnancy becomes an ethical issue, whether or not the body in question is caught up in explicitly controversial practices (such as surrogacy, abortion, reproductive technology). Even in 'normal' pregnancy, what the mother does with her body, what she ingests, how she moves and when, is considered to be a legitimate target for moral concern. Hence there are a myriad of discourses on the right way to get pregnant and stay pregnant. Without denying that the concern about reproductive practices discussed in chapter 1 is motivated by a genuine consideration for the health and wellbeing of mothers and their children, that concern is not outside those mechanisms which attend to the reproduction of the body politic and its asymmetrical power relations. This well-meaning surveillance is also consistent with political investment in 'healthy', productive bodies suitable for exchange in a labour market.

The differential treatment the maternal body receives illustrates how contemporary techniques of self-formation constitute and allow for differences. But this is not necessarily to the benefit of women. While the maternal body is the target for apparently unlimited surveillance, it does not get absorbed into the labour market it reproduces. As Moira Gatens argues, from the time of its inception the social contract has implicitly or explicitly excluded the labour of women's bodies from the realm of socio-political freedom (Gatens 1991a: 34–44). Woman's labour has been privatised and the product of her labour has traditionally been viewed as belonging to someone else. The labour of pregnancy has also been privatised but remains, paradoxically, open to public scrutiny in the interests of the health and welfare of the social body. Of course, feminism has attempted with some success to extend the principles of 'free contract and exchange' to women, and in this context the apparent proliferation and commercialisation of surrogacy and other reproductive practices should not be a surprising by-product. But while women's labour is no longer explicitly excluded from the body politic, the labour of pregnancy and the maternal ethos remain foreign to the social contract as I have indicated. Those who would seek to ban surrogacy and reproductive technology (feminists included) perpetuate this exclusion of the maternal body from self-motivated social exchange. And in assuming a space for pregnant women outside of patriarchal domination these critics miss, and inadvertently participate in, the work of more insidious modes of subjection and normalisation operating under the guise of care and the promotion of freedom.

That the techniques of power which Foucault describes reproduce sexual difference to the disadvantage of women is well substantiated by the instance of the constitution of a maternal ethos. But the pregnant body is an obvious mark of sexual difference and the differential treatment it receives may not apply to women generally. So let me consider for a moment the significance and constitution of other kinds of female bodies – those who now occupy public habitats along side men. To the extent to which women have been admitted to the institutions Foucault discusses, our habitats have been altered, and with this we would expect, if we follow his model of disciplinary power, attendant changes in embodiment. Yet even if we grant that women undergo the same techniques of discipline as men in preparation for entry into the labour force, there is much to suggest that these techniques do not obliterate sexual difference or consequent inequitable ranking. Traditional concepts of femininity such as those of mother and whore, work together with new images of femininity and ideals of female independence to produce paradoxical habitats for women. Contradictions arise: the female worker must act like a man as she crosses her legs, masquerades as a lady and takes responsibility for fraternising with the men. These contradictions constitute a female ethos which not only guarantees the reproduction of sexual difference but often debilitates women. As the work of Susan Bordo indicates, all too often our bodies, in assuming an ethos of resistance to patriarchal power, wear the marks of self-negation or, in the extreme, hysteria, agoraphobia and anorexia nervosa (Bordo 1989).[9] This is not to suggest a return to more conventional habitats or that these are free from contradiction. Rather it suggests that 'egalitarian' discourses and practices do not produce equality or sameness between the sexes: it would seem that the body politic needs sexual difference (for reasons I'll get to) and that insofar as techniques of discipline and surveillance reproduce asymmetrical power relations, then this also applies to relations between men and women.

AN AESTHETICS OF SEXUAL DIFFERENCE?

So far I have been discussing Foucault's model of how contemporary discourses and practices constitute the individual's embodied ethos and the ways in which this model challenges universalist and contractarian ethics. Once it is suggested that the individual's 'moral identity' is not given but socially constituted through a normalising relation to a moral code, then ethics may be found to be unethical insofar as it reproduces inequalities. I have also suggested that, in the absence of any alternative account, Foucault's model of embodied self-formation poses difficulties for an ethics of difference which focuses on prohibitive operations of power and which assumes either the ability to escape power relations altogether or the ability to abstract oneself from one's embodied ethos in an interactive dialogue with the other. Finally, I have suggested that, with some work, Foucault's model of the normalising, constitutive relation between a disciplinary moral code and an embodied ethos is useful in accounting for the social reproduction of sexual difference and its asymmetrical power relations.

I mentioned at the beginning of this chapter that an ethics of sexual difference requires some account of the genesis of identity and difference: what is the status of the other's difference, how is it generated and how does it relate to my identity, if at all? How the generation of identity and difference is understood will determine what an ethics of difference may look like and whether it can fairly account for sexual difference. Foucault's analysis of disciplinary power does include, as discussed, some account of how differences are constituted and reproduced within a hierarchy of domination and control. And his analysis of confession points to the constitutive and normalising effects operating in the self-other relation (at least in its institutionalised forms). But, as I will go on to argue, his account of the production of differences is insufficient for an ethics of difference in general, and an ethics of sexual difference in particular. The inadequacy becomes apparent when exploring Foucault's own ethics of difference: his aesthetics of self-formation.

Foucault objects to the normalising effects of contemporary modes of self-formation – effects that exclude and denigrate certain bodies, experiences and sexual practices. As techniques of self-knowledge and confession are involved in this normalising operation of power, it is not possible to successfully oppose domination or exclusion by affirming some as yet undiscovered truth of oneself. He suggests, then, that 'the target nowadays is not to discover what we are, but to refuse what we are. . . . We have to promote new forms of subjectivity through the refusal of this kind of individuality' (Foucault 1982: 216). And there is some agreement within feminism about the strategy of refusal:

> If she confines herself to asking the question of woman (what is woman?), she might merely be attempting to provide an answer to the honourable male question: what does woman want? She herself still remains the *object* of the question . . .
>
> The gesture that the 'historical moment' requires might be to ask the 'question of man' in a special way – what is man that the itinerary of his desire creates such a text? . . . [otherwise we] will continue to preserve masculinity's business as usual and produce answers that will describe themselves, with cruel if unselfconscious irony, as 'total womanhood'.
>
> (Spivak 1982: 185–6)

For Gayatri Spivak, refusing the question 'what is woman' involves turning the question against the listener, indirectly. Foucault's refusal involves the promotion of new modes of subjectivity – a strategy which requires further examination.

To suggest that it is possible to re-create oneself differently seems to imply again a space for agency outside of the effects of a disciplinary moral code and it seems to assume a locus of difference beyond the kind of difference constituted by and denigrated within the hierarchical spaces of discipline and surveillance. Minimally it seems to imply a precultural body which lies in waiting for more creative techniques of self than those currently in operation. But perhaps not. While subjection is the subjection of bodies, the promotion of new modes of subjectivity cannot simply involve asserting mind over matter. According to Foucault, the operation of power is ahead of conscious intervention:

> What I want to show is how power relations can materially penetrate the body in depth, without depending even on the mediation of the subject's own representations. If power takes hold on the body, this isn't through its having first to be interiorised in people's consciousness.
>
> (Foucault 1980: 186)

Given the pervasive operation of normalising modes of subjection, it is difficult to imagine how and from where new modes of subjectivity could emerge. Foucault's belated interest in Ancient Greek sexual ethics is part of a search for ways around this apparent impasse.

He finds a certain flexibility operating in Ancient Greek ethics, particularly in regard to the two factors involved in the constitution of one's ethos which I have already discussed: the relation to a moral code and one's self-forming activity. The way in which the individual establishes his relation to the moral code in order to enjoy his pleasure as he 'ought' is the mode of subjection. In contrast to Christian ethics, the Greeks did not tend to 'trace out the boundary of the prohibitions' with regard to sexual practices (Foucault 1987: 58). Rather, their 'use of pleasure' allowed the individual to manage and adjust his own bodily activities according to his needs, the time and place and his status. Therefore, to quote Foucault, 'in this form of morality, the individual did not make himself into an ethical subject by universalizing the principles that informed his action' (ibid.: 62). It is the absence of a universal moral code and the normalisation this implies which makes Greek ethics appealing to Foucault.

What is also attractive to Foucault is the way the self-forming activity of Greek ethics avoids the trap of the truth effects of power. The self-forming activity of Christian ethics is the kind he rejects: the hermeneutical practice of self-reflection and interpretation, combined with purification and renunciation. In Ancient Greek thought, he claims, this activity is in the form of domination, not in the presence of some external judge or agency of domination, but of one part of the self (Reason) over another (appetite). 'Ethical conduct in matters of pleasure was contingent on a battle for power' (Foucault 1987: 66). But this was a battle within the self; it involved *control* and transformation of desires rather than their suppression and was supported by affiliated techniques of self-transformation such as physical exercise. Virtue was not a question of self-knowledge and confession but of active self-mastery and moderation.

Despite this apparent tolerance of different modes of being in Greek ethics, Foucault does note some elitist and oppressive aspects. In a rare acknowledgment of the asymmetrical relation between the sexes, he observes that women pay a high price for the active freedom enjoyed by men in the practice of an ethics of self. He gives two reasons for the disqualification of women from the position of ethical subject in Greek thought. The first is that a woman's role as wife and mother meant that her use of pleasure was not self-designated: 'moderation was not self regulated but *imposed* on them by their condition of dependence in relation to their families' (Foucault 1987: 82). Secondly, Foucault points to a structural problem: as moderation has an essentially masculine structure of active virility, 'immoderation derives from a passivity that relates it to femininity' and hence women (ibid.: 84).[10] Just when Foucault finds evidence of an aesthetics of self he also finds systematic exclusions. The question remains: are the two connected, and if so, why?

In an interview where Foucault reviews *The Use of Pleasure*, he emphasises his disapproval of the exclusions that this Greek ethic effected. And he poses the question:

> Are we able to have an ethics of acts and their pleasures which would be able to take into account the pleasure of the other? Is the pleasure of the other something which can be integrated in our pleasure, without reference either to law, to marriage, to I don't know what?
>
> (Foucault 1984: 346)

Without seeking solutions to the present in the past, Foucault believes such an ethics of difference is possible. Using the idea of the body as aesthetic material, he suggests that it is possible to 'create ourselves as a work of art' (ibid.: 351). By working on our bodies, transforming our habits and expanding our capacities we can create ourselves differently without reference to the normalising disciplinary structure and without domination and exclusion of the other. This explicitly Nietzschean 'aesthetics of existence' requires, according to Foucault, rejecting the 'idea of an analytical or necessary link between ethics and other social, economic or political structures' (ibid.: 350). In other words these techniques of self, this ethics of difference, lie outside the social, economic and political institutions which disseminate disciplinary power.

Without dismissing Foucault's invaluable analytics of power and his elaboration of the normalising constitution of an embodied ethos, I find this aesthetics of self-formation limited in its potential. It would be fair to suggest that women would remain sceptical of basing an ethics of difference on creating the self as a work of art. This scepticism comes in part from the suspicion that a certain kind of male body is already a work of art so that even the most flexible communal ethos is reluctant to value or recognise women's embodied modes of being. Moira Gatens, for example, has claimed that our body politic speaks only of one body, one reason, one ethos – one which is isomorphic with the male body (Gatens 1991b: 81). As I have argued in my account of the ethics of reproduction, the maternal body is one kind of body not accommodated in the body politic. Foucault would agree that an assumption of sameness has excluded the female body among others from representation within the tissue of the body politic – indeed he acknowledges this in his discussion of Greek ethics. But he does not seem to appreciate the effects of these exclusions upon the possibility of an ethics of sexual difference. To suggest that those women and others, already excluded from social exchange, should practice an aesthetics of self outside existing social, economic and political structures would perpetuate their exclusion. I am not suggesting that women are incapable of practising an aesthetics of self: that women now occupy places in the public sphere suggests that women can and do reconstitute themselves through a change in habits and can accrue some social value accordingly. But, as I have argued, even given this work, women are far from being at home with themselves or with the public body. If these difficulties are an effect of the privileged position given the white, middle-class male body, then something, it seems, has been left out of Foucault's model of self-formation – an account of the generation of value in the production of identity and difference.

Just as no ethos is given outside of social discourses and practices, no work of art is created in a vacuum. While Foucault derives his notion of creating the self as a work of art from Nietzsche's philosophy, he omits Nietzsche's analysis of how the artefact obtains its meaning and value. To quote Nietzsche: 'whatever exists, having somehow come into being, is again and again interpreted to new ends' (Nietzsche 1969: 77) and 'only that with no history is definable' (ibid.: 80). The history by which the self attains a definition, an identity and value is, according to Nietzsche, through the measurement of the self against the other (ibid.). This involves evaluation and interpretation of the other's difference according to socially specific concepts. I will be discussing Nietzsche's ethics of difference in more detail in chapter 5. The point I wish to raise here is that the value attributed to any body as 'a work of art' is generated through comparison with, and often denigration of, the other.

It is necessary to consider not only the way in which social values become embodied in an ethos, but also the role of the self–other relation in this process of production. Ethics of care has already put the self–other relation on the ethical map by legitimating the kinds of moral values relevant to its maintenance: love, care and friendship. While important, such a focus assumes that the self–other relation is already in place and that the identity and difference of the parties is given prior to care. With identity and difference given, what matters in an ethics of care is that the other's difference be considered in one's resolution of moral dilemmas. I am suggesting that it is also necessary to consider the possibility that the other's identity and difference and the value I attribute to them may be reconstituted through my evaluation of them. It is the constitution and reformation of the identity, meaning and value of each party in relation to the other which requires further attention.

Foucault seems to forget the history of self-formation whereby the value of one's ethos (aesthetic or not) is generated through one's relations with others. He does acknowledge that, in Ancient Greek ethics, besides a relation to his body, the constitution of a man's ethos did require some relation to others. The two kinds of relations involved were a man's economic relation to his wife and an erotic relation to boys (Foucault 1987: 152–225). I have already mentioned how the ethical subject's relation to his wife was such that she, and women in general, were excluded from practising an aesthetics of self. The relation which seemed to matter in Greek techniques of self-formation was a man's erotic relation to boys. In the *Symposium*, for example, Plato sets out the way in which a man's contemplation of the beauty in the body of the male object of desire is transformed into contemplation of the Form of Beauty (Plato 1951: 92–4). This partnership with a boy, and the movement of desire within it from physical to 'spiritual' in the presence of a man's body-form, was central to a man's self-transformation into an ethical, knowing subject. So an aesthetics of existence does require a relation to another – at least it did for the Greeks.

Foucault's interest in this relation is to stress not so much the necessary presence of the other but the tolerance of same-sex relations in Greek ethics: as moderation was what mattered, rather than rules about the sex of one's object, techniques of self-formation were various rather than universal. But, as Rosi Braidotti notes, while Foucault acknowledges the presence of another in the constitution of the self, he seems blind to the consequences of the fact that this 'other' was male (Braidotti 1991: 91). She argues more generally that, whether addressing the subjection of bodies through disciplinary power or an aesthetics of self, Foucault is 'speaking about the man's body' (ibid.: 95). To Braidotti's point

I would add that Foucault also seems blind to how the status, form and value of this man's other (whether male or female) contributes to his own identity and value. The concern Foucault highlights about a man's relation to boys in Greek ethics is the anxiety its asymmetrical structure generated: a boy could not identify with his position as the passive (feminine) object of a man's desire if he was to later become an active, free agent (Foucault 1987: 221). Foucault does not seem to notice that the asymmetry here, and the designation of passivity as feminine, may constitute the privileged status of the ethical subject. Nor does he appreciate that the exclusion of women may generate the value attributed to the male body as a work of art.

Even in Foucault's earlier work on disciplinary power, there is a blindness to the role a relation to the other plays in the generation of identity and value. In *Discipline and Punish,* for example, he provides no account of how the time–space segments of individualisation gain their code and rank. The meaning and value of these institutional habitats determine, according to Foucault, the ethos of those who occupy them. Insofar as these spaces are different in code and rank, they produce differences in both the identity and status of the occupants. Yet the value and meaning of these habitats is not given outside of their relational context, nor is their significance static. An account of the production of meaning and value, identity and difference would show how the spaces of discipline and surveillance in the workplace, for example, are governed by oppositions such as clean/dirty, managerial/manual. The code and rank of each space is determined by the *relation between* these oppositions. A managerial space, for example, accrues more value (and less surveillance) not because of any inherent value but because of its relation to and difference from the spaces of manual labour. Second, and perhaps more important for our purposes, the code and rank of Foucault's spaces of individualisation are determined by the relative value of the bodies occupying these spaces at any particular time. So when men enter a work space traditionally occupied by women the value of that space will increase and the degree and nature of surveillance and discipline will alter accordingly. When women occupy spaces traditionally the home of men they become enigmatic: their bodies are 'masculinised' and the space 'feminised'.[11] Nothing inherent about the bodies of those who occupy these habitats determines their value and status. Rather, it is the difference between them that counts.

It is not enough, then, to acknowledge that women have traditionally been excluded from the body politic. It is also necessary to trace how the male body, as a work of art, has derived its value from this exclusion and from the devaluation of women's manner of being. As Adrienne Rich suggests:

> Women's honor, something altogether else: virginity, chastity, fidelity to a husband. Honesty in women has not been considered important. We have been depicted as generically whimsical, deceitful, subtle, vacillating. And we have been rewarded for lying . . .
>
> Lying is done with words, and also with silence.
>
> (Rich 1980: 186)

Woman's deceit and silence is what gives her a place of sorts within the social fabric. And this place holds something together – the carpet of truth, an ethic between men. To quote Rich again, 'The pattern of the carpet is a surface. When we look closely, or when we become weavers, we learn of the tiny multiple threads unseen in the overall pattern, the knots on the underside of the carpet' (ibid.: 187). Insofar as women accrue social value as women it is through an ethos which upholds an ethic of sameness among men. While the effect of women's work is on display, the labour of women itself is unseen, unrepresented in the pattern of social life. If a male 'ethos' is upheld by woman's silence then this suggests that this ethos is not self-contained – its value is generated through its relation to and difference from others who are excluded as a consequence. Nor does the inclusion of others, without also interrogating the process of production of value, solve the problem: if the dominant public ethos gains its meaning and value from the exclusion of others then it is little wonder that the inclusion of these others makes little sense at times – pregnant bodies being a case in point.

Evoking an aesthetics of self which is practised apart from others and outside social institutions disavows, rather than avoids, this process of production of value involving the denigration of others. The generation of identity and value by the denigration and exclusion of others is apparent in Ancient Greek ethics and, without any indication to the contrary, would also be true of the production of Foucault's contemporary corporeal artefact: the self as a work of art, its meaning and value, is not given outside its relation to others even if it is built without explicit reference to a universalising moral code.

Given the exclusions already in place, an 'aesthetics of existence', while appealing, is an option more open to white middle-class men than it is to women. The proposal that this ethics be practised apart from 'other social economic and political structures' marks the point where a feminist ethics of difference should part company with Foucault. The value attributed to any body is generated precisely through these other structures and within a field of relations with others in such a way as to favour the production of a specific kind of male body. Therefore, while women may have more choice today in the 'art of living', we are also aware of how our self-forming activities are devalued and reappropriated in ways contrary to how we apprehend them. Feminism cannot afford to separate the ethical relation to the self from the operation of other social structures. But the need to attend to both does not exclude the possibility of change, as Foucault suggests (Foucault 1984: 350). On the contrary, only through an interrogation of the assumptions regarding the sexed body that maintain these other structures, and hence men's privileged position, is an aesthetics of existence possible for women at all.

While escape from the other and from social institutions is not possible, change is. It is possible, on the one hand, to hold that one's ethos is embodied and socially constituted and that this identity is generated through differential relations to others, and on the other hand to claim there is a remainder of difference, or a space for agency, which works against this operation of power. This space for change does not depend on positing an agent or a body prior to or outside of the workings of social discourses and practices. Recourse to a pre-social subject or a pre-social body to establish an ethics of difference falsely assumes, as Judith Butler claims, that:

> (a) agency can only be established through recourse to a pre-discursive 'I', even if that 'I' is found in the midst of pre-discursive convergence, and (b) that to be *constituted* by discourse is to be *determined* by discourse, where determination forecloses the possibility of agency.
>
> (Butler 1990: 143)

One's ethos can be said to be constituted but not determined by cultural discourses and practices because of a slippage operating in the self–other relation. As Butler goes on to suggest, there is an 'excess that necessarily accompanies' the construction of the self in and through the other where the identity so constituted is said to be autonomous. It is this excess which is the condition of agency and change (ibid.: 143, 147).

Butler's crucial claim about an excess of difference in the self–other relation, while evoking a long philosophical tradition about the constitution of identity and difference, is hardly heard in Anglophone philosophy and even more rarely understood. For that reason I will spend some time in the next two chapters elaborating this thematic. The point I wish to draw from it now is that both women and men can reconstitute themselves through a change of habits, not because their embodied identity lies outside culture or outside relations to others, but because identity is produced in relation such that an ethos is always deferred towards other possibilities. Feminists share Foucault's desire to open up these other possibilities. But while he would do it through working on the self in apparent isolation from others and social norms, a feminist ethics would do it through a critique of those social discourses which both constitute women as 'other' to men and deny any other possibilities.

The work of Luce Irigaray is exemplary in its interrogation of the means by which identity and difference are generated in the cultural production of sexual difference. In contrast to the contract model of social relations discussed in chapter 1,[12] where it is assumed that the identity of different individuals is given prior to their relations, Irigaray holds the view that identity and difference are constituted in relation. Moreover she claims that the constitution of sexed identity and difference is such that women are excluded from the position of the subject of discourse and social exchange. Rather than lamenting this exclusion and/or affirming a feminine mode of being outside of the forms of social production which effect it, she examines the conditions which give rise to this exclusion. Such an examination requires a revolution in ethics where ethics is understood as the problematic of identity or place (Irigaray 1991: 166, 167).

What must be examined in an ethics of sexual difference, according to Irigaray, is the 'economy of the *interval*' between the (male) subject and discourse, between the subject and his world and between the subject and woman (Irigaray 1991: 166, 169). The 'interval' is the distance or difference created between the subject and his others such that he can claim autonomous self-identity. This interval is *created*, not given. What Irigaray finds within the economy which creates the interval between man and woman is this: in order that man can constitute his place in the world, woman is denied a place of her own. Particularly in her role as the mother, 'woman represents the place of man'; she is the 'envelope' by which man delimits himself and, as the mother and man's other, 'woman remains the *place separated from its "own" place*' (ibid.: 169). Another way to put this is that sexual 'difference' is constituted in such a way that man gains his autonomous identity at the expense of woman. Irigaray's examination of the constitution of sexed identity and difference exposes a structural reason why women's modes of being, particularly motherhood, continue to be excluded from the position of subject of social exchange.

Women not only lack a place of their own in this economy of difference but, Irigaray notes, women's 'difference' is understood and constituted in relation to man's identity – as lacking what man is said to have (Irigaray 1985: 69, 74–5). The constitution of identity and difference is such that there is no sexual difference to speak of: only sameness or a lack. Irigaray wants to do more than examine the conditions which give rise to man's identity. She also wants sexual difference to take place: she wants to create the possibility of an ethos for women. Now Irigaray, like Foucault, recognises that one's place or ethos is constituted in and through the body. So, if woman's difference is constituted in relation to man's identity, and if woman's 'difference' so constituted is embodied, then, to return to the question raised earlier, how and from where can another sexual difference emerge which will give women a better deal? This different kind of sexual difference emerges, not from outside the ethos of woman constituted as man's 'other', but from the 'images' of woman already deposited within the conditions of production of sexual difference (Irigaray 1991: 169). That is, new modes of being for women emerge from this interrogation of the economy of the interval itself. This reconstitution of female habitats occurs 'between what is already identified' (ibid.: 168), between those poles of sexual difference we take for granted. This ethics of sexual difference is a material practice (an opening of one's embodied ethos towards other possibilities); its products emerge from challenging what already is; and the character of the emerging modalities of existence cannot be predicted (ibid.: 174, 169, 176). This possibility again depends on the excess Butler describes: the excess of difference operating in the accepted economy of sexual difference; it depends upon, as Irigaray puts it, the '*residue* of any creation or work' (Irigaray 1991: 168).

I have moved too quickly over this ethics of sexual difference and propose to now retrace some of its steps. Rather than provide a commentary of Irigaray's ethics, which would take us into debates about whether or not her work is essentialist and/or anti-feminist,[13] I shall present a genealogy of sexual difference. And rather than claiming that the economy of sexual difference is such that a residue, remainder or excess of difference exists from which new modes of being may emerge, I propose to demonstrate this by interrogating the economy itself.

I begin with Hegel because, contrary to contractarian theorists, he takes sexual difference seriously. He does not suppress the interval between self and other but puts it on the map. That is, Hegel, as I have mentioned, does not believe that self-identity is given prior to relations with others. Rather, he suggests that identity and difference are constituted through that relation, through the production of an interval between self and other, between man and woman. I will attempt to show, through my rereading of Hegel, that in recognising this economy Hegel exposes a debt man owes to woman for his privileged position. On the other hand, Hegel does not seem to recognise a residue or excess of difference operating in this economy. He proposes instead a unity between self and other, a unity of identity and difference. Insofar as he insists on this unity Hegel effaces difference to the disadvantage of women. I will argue that such a unity is impossible if one pays attention to the social constitution of embodiment within the economy of identity and difference.

Chapter 3

Hegel's restricted economy of difference

In the previous chapter I redefined ethics, following Foucault and Irigaray, as the examination and practice of that which constitutes our embodied place in the world. In this chapter I will extend that definition through an exploration of Hegel's ethics, in particular his theses that identity is constituted in relation to what is different and that a unity of sexual difference is realised in ethical life.

I have suggested that an ethics of sexual difference needs to account for how identity and difference are generated in and through the other. Or, following Irigaray, it is necessary to pay attention to how social discourses and practices constitute the interval or difference between man and woman. There are two indications of what might be found through such an interrogation. First, insofar as a certain kind of male ethos is privileged in our society, its meaning and primary status is generated by constituting women's ethos as 'other' and by excluding this ethos from the benefits of social exchange. Moreover, to maintain this male ethos as the norm, its identity and value is said to be given in isolation from others: the production of the interval between self and other is suppressed and denied and hence so is the debt men owe to women for their privileged status. Second, there is an excess of difference lurking in the production of the dominant male ethos. This excess or residue is that from which various aesthetics of existence for women may emerge. This other difference is not outside the conditions which privilege man and devalue or idolise woman as man's other but, rather, emerges from an examination of those conditions.

In this chapter I will begin to elaborate both these indications about the genesis of sexed identity and difference. I have chosen to do this via Hegel's philosophy for two reasons. First, the idea, being rediscovered in contemporary ethics, that a person's identity is produced through their relations with others can be traced back to Hegel's philosophy. Second, Hegel's ethics reflects some sentiments (and problems) expressed in contemporary ethics of difference. Despite taking difference seriously, Hegel views human progress in general, and the development of rational subjectivity in particular, as an overcoming of differences. For him there is an original and final unity of identity and difference: the subject strives to make what is different (the other and the world) a reflection of its own identity. So, while elaborating Hegel's claim that identity is generated through the other, I will also examine the conditions necessary for the social unity he proposes. With particular reference to Hegel's thoughts on ethics and sexual difference, what will be exposed is that the unity of identity and difference is impossible. Within Hegel's own terms there will always be an excess operating which mitigates against unity and sameness. On the other hand, to the extent that Hegel insists on this unity of differences he does so to the detriment of women.

As the kind of social harmony and overcoming of differences Hegel proposes is said to occur through a dialectical relation between the universal and the particular, between the community and the individual and between the self and others, the terms are the same as those proposed within Porter's ethics of synthesis (1991), mentioned in chapter 1. Porter, unlike Hegel, wants to preserve differences against the onslaught of universalism. Yet as her terms are the same as Hegel's, the outcome, I suggest, would also be the effacement of differences, despite her best intentions. Similarly, as Hegel's proposal for the unity of identity and difference relies on the same kind of rationality and self-transparency as Benhabib's 'interactive' ethics, discussed in chapter 1, then the indications are that this ethics would also efface the differences it seeks to accommodate.

The question I address in the following reading of Hegel's philosophy is this: if sexed identity is constituted in and through one's relations with others (as Hegel, Porter and Benhabib suggest) then what is the origin and status of the differences which Porter and Benhabib seek to preserve but which they may, along with Hegel, inadvertently efface? Benhabib's discussions of Butler and Hegel are instructive here. On the one hand, Benhabib puzzles over Butler's claims, which I raised in chapter 2, that the subject's identity is divided and that, while constituted by discourse, this identity is not determined by discourse (Benhabib 1992: 215–18). These claims, says Benhabib, are not sufficient to account for agency and change (while her own interactive rationality is). On the other hand, in her own astute reading of Hegel on sexual difference, Benhabib finds, and in fact concludes her book with, evidence which can only support Butler's claims. Hegel, in ways I will go on to outline, is the father of the thesis that the subject's identity is divided (between itself and the other) and that identity is determined by discourse (insofar as Spirit is objectified). Part of his progression towards the unity of identity and difference involves the dissolution of sexual difference. But, as Benhabib shows, such a 'unity' between the sexes can only be achieved by expelling women from public life (ibid.: 255) and, even with this done, 'what remains of this dialectic is what Hegel precisely thought he could dispense with: irony tragedy and contingency' (ibid.: 256). Irony, as I will go on to show, is 'the otherness of the other', that incalculable residue of difference which cannot be subsumed under any universal. Part of my purpose in rereading Hegel's ethics is to argue, in support of Butler, that this 'irony' or difference is the locus of agency and the possibility of social change. From it can emerge open-ended possibilities for women, not through a synthesis of opposites nor through interactive rationality alone but through an interrogation of the conditions which constitute and suppress it to the disadvantage of women.

One final introductory comment: in turning to Hegel's thesis about the genesis of identity and difference, I am not abandoning the problem of the social constitution of embodiment. On the contrary, while Hegel is more usually read as a philosopher of consciousness, he does have an account of how one's ethos is constituted through the body. He is also aware that the body is the locus of one's ethical encounter with others. This body, with its actions, habits and capacities, is the first sign of the self. Paying attention to this aspect of his philosophy not only yields a more thorough account of the social constitution of sexed identity and difference, but insofar as Hegel entertains a unity of identity and difference it is based, I will argue, on the reproduction of a particular kind of male body.

IDENTITY, DIFFERENCE AND THE HABITUAL BODY

I mentioned that the idea of divided subjectivity can be traced back to Hegel. There are two ways in which the subject is divided. First, the subject is divided in terms of temporality in that, as Hegel demonstrates in his 'dialectic of meaning', what is present or 'now' is so by virtue of its difference from the past and the future (Hegel 1977: 60). Self-present identity is divided between what is present and what is not. Second, the subject is divided spatially between itself and an outside which it negates in order to be. The subject's identity is divided between the 'here' and the 'there', between itself and what it is not. Or, as Hegel suggests in more concrete terms in his 'master/slave' dialectic, the subject's identity depends on positing a difference from his other and upon the other's recognition of this identity (ibid.: 111).

By linking these aspects of divided identity, Hegel can claim, more generally, and against contractarian ethics and liberal individualism, that identity which claims to exclude all difference is merely an abstract self-relation (Hegel 1975: 166–7). While identity is self-relation, it contains 'essentially the characteristic of *Difference*' (ibid.: 168). No entity is autonomous or simply self-present and self-identical; it has within it its negative, that is, reference to another or mediation. As there is no identity without difference, the subject is torn between himself and his other.

While Hegel exposes the debt Identity owes to Difference, he cannot entertain the possibility of differences not in the service of a universal identity. Beneath differences there is a deeper identity, the ground or the unity of identity and difference, towards which the dialectic proceeds. It is through his notion of *Geist* (Spirit) that Hegel attempts to reconcile identity and difference (or the necessity of autonomy with the possibility of unity within and between individuals).

Spirit is not above and beyond particular different individuals, nor is it something that individuals bring about. Rather, it is the movement whereby difference (as opposition) proceeds from and returns to identity so that unity (or Substance as Subject) is:

> the movement of positing itself, or is the mediation of its self-othering with itself . . . [I]t is the doubling which sets up opposition, and then again the negation of this indifferent diversity and of its antithesis. Only this self-*restoring* sameness, or this reflection in otherness within itself – not an *original* or *immediate* unity as such – is the True. It is the process of its own becoming, the circle that presupposes its end as its goal, having its end also as its beginning; and only by being worked out to its end, is it actual.
>
> (Hegel 1977: 10)

Individual differences, including sexual difference, are moments of this movement – moments of Spirit's self-expression as a particular embodiment. While differences arise out of unity, oppositions are subsequently overcome in the inevitable path to a higher reconciliation. The reconciliation towards which Hegel's philosophy aspires is where individual self-consciousness is represented in social reality, that is, where Spirit is objectified such that the 'I' is 'we' and the 'we' is 'I' (Hegel 1977: 110).

This unity is Hegel's formula of ethical life where all individual differences including differences between the sexes are dissolved into a harmonious community life. In the *Phenomenology of Spirit,* he uses the story of Antigone to show how the conflicting actions of men and women (insofar as they are ethical) are partial expressions of, and necessarily lead to, a social unity. For Hegel, *Antigone* is a picture of ethical life where individual moral autonomy is a moment of difference within a self-restoring sameness (Hegel 1977: 10, 260–1).

Before assessing Hegel's account of ethics and sexual difference it is necessary to retrace the process by which community consciousness supposedly comes to reflect the will of all individuals. This is a two-way process: the individual's ethos is constituted to reflect community ideals, and individual will, together with others, reconstitutes the social world to reflect itself. The issue to be examined is the extent to which, given ideal circumstances, an individual ethos can be said to represent and reflect that of the community and vice versa such that differences are accommodated within communal institutions and practices. The issue can be reformulated in terms of Hegel's theory of the sign.

Hegel's claim that difference proceeds from and returns to a unified sameness is grounded in, yet undermined by, his theory of the sign.[1] In his discussion of language he claims that meaning participates in the universal and is always embodied in a sign in such a way that the body of the sign (its material signifier) is transformed to represent a meaning other than itself (Hegel 1971: 212–13). That is, words and actions are transformed to represent a meaning given by conscious ideas and intentions. But ideas and intentions are not isolated, self-identical units of meaning. As I have mentioned, the meaning of a particular sign is never simply present, nor therefore is the conscious idea which informs the sign. Rather, meaning is produced differentially by the sign's relation to and difference from what it is not.[2] The importance of Hegel's semiology resides in his analysis of the material effects of signification (how material signifiers are transformed within a social context in order to mean anything at all) and the alterity and instability inherent in the production of meaning.

Left at that, Hegel's semiology would indicate that the individual's ethos is transformed into a sign of community ideals without the possibility of being reduced to those ideals. But Hegel also attempts to close off his own insights by reducing the sign to a transparent signifier of conscious, rational thought. He proceeds as if material signifiers (words, actions, etc.) already anticipate their idealisation in the sign, as if individual thought and universal meaning come before signification and as if the body of the sign submits without remainder to the meaning-creating activity of the mind. Given this reduction, Hegel will assume that an individual's ethos (their actions etc.) can be transformed into a direct representation of the individual's will which in turn has been transformed into a transparent sign of universal will. But, as I will argue, this reduction of an individual ethos to a sign of rational consciousness remains incoherent, in Hegel's own terms.[3]

While Hegel is usually read as a philosopher of consciousness, it is not the case that the will of the community works directly on the will of the individual. Individual will can only become a moment of universal (community) will via the transformation of a material signifier. And the most immediate sign of the will is the body, its actions, capacities and so on. In terms of the following reading of Hegel, the body is moulded to represent universal will through the mechanism of habit formation. Further, individual will is not pre-social. It does not arise before habit formation but is an uncertain effect of that process. That individual will is an effect, rather than cause, of embodied modes of being, has a bearing on the success of Hegel's proposal to unify all differences.

Hegel begins his account of the production and subjection of individual will in the *Philosophy of Mind* with the claim that the Soul (Spirit or the 'we' that is 'I') is the 'basis of all . . . individualising of mind' (Hegel 1971: 29). But the 'I' does not emerge directly from community will: individual consciousness arises only as an effect of the 'we' passing through the body; individual will is an effect of the social constitution of the body. The relation between the social will and the individual body is not a copula between two entities which pre-exist that relation. As with the unit of representation in language, the material component, in this case the body, is transformed in order to signify universal consciousness.

> The Soul, when its corporeity has been moulded and made thoroughly its own, finds itself there in a *single* subject; and the corporeity is an externality which stands as a predicate, in being related to which, it is related to itself. This externality, in other words, represents not itself but the soul, of which it is a *sign*.
>
> (Hegel 1971: 147)

The body is the community's work of art and mode of expression. And in representing the will of all, the body, like any material signifier, becomes other than itself (ibid.).

According to Hegel, individual consciousness (and therefore individual will) comes into being or 'awakens' as it 'distinguishes itself from its *mere* being' (Hegel 1971: 65), that is, from the body by which it is immersed in life. Hegel implies that this 'awakening' is spontaneous. However, he also suggests that it is a function of the body being motivated by the Other (Hegel 1971: 55–64). The Other, who represents social consciousness, objectifies the child's body in ways consistent with community values. It it through this objectification, occuring in the process of caring for the child, that a distinction between 'inside' and 'outside' the body arises and from this a distinction between self and body. Taking on the distinction between self and body is the first moment of intelligence, one which manifests as a difference in modes of affection: the child has *sensations* (ibid.: 73), which appear to arise solely in the body, and *feelings*, which seem to originate in the mind (ibid.: 75). The particular embodied self begins to emerge from the Other when sensations and feelings are unified. Bodily sensations are internalised by being claimed as one's own and (mental) feelings are embodied. The subsequent unity is the feeling, sentient and particular self (ibid.: 92–3). But this primordial self is not yet a conscious agent: it is still immersed in corporeality, it has not distinguished itself properly from an outside and is actualised and marked by others (ibid.: 94–5, 101–4). The Other (mother) is still the subject of the child, the two being interrelated by a 'sensitivity' or 'rapport' by which the child is subjected.

The pre-subject becomes a conscious subject proper when it can distinguish itself both from the immediacy of its embodiment (that is, from itself as a sentient self) and from its Other. On the distinction between the individual self and its embodiment, Hegel claims that:

> This particular being of the soul is the factor of its corporeity; here we have it breaking with this corporeity, distinguishing it from itself . . .
> And consciousness it becomes, when the corporeity, of which it is the subjective substance, and which still continues to exist, and that as a barrier for it, has been absorbed by it, and it has been invested with the character of self-centred subject.
>
> (Hegel 1971: 140)

Hegel claims that this distinction between the self-centred subject (who is a moment of soul or community spirit) and its particular bodily self arises spontaneously from 'the contradiction of being an *individual*, a *singular*, and yet being at the same time immediately identical with the *universal*' (Hegel 1971: 125, 139). This refers to Hegel's claim that any entity comes into being in relation to a social whole which can absorb it without remainder. Yet his description of the sentient self suggests that it is neither a unified singular nor universal. There is already an excess to this simple unity – that which already allows a '*rapport*' with, and subjection by, others (ibid.: 104). In other words, individual consciousness is not a result of a spontaneous differentiation of community consciousness. It is a function of the body being constituted and directed by others in a social context. Consciousness is the process of making one's body an object for oneself in response to being objectified by others. The question is: what happens to the rapport between self and Other in this process?

The demand of a sociality which assumes difference as a moment of community ethos is to remove the immediacy of corporality and transform the self into a unity free from, but the same as, its Other (Hegel 1971: 125). The construction of the proper body involves translating the motivating activity of the Other into a self-relation (transforming control by another into self-control), and this occurs through the mechanism of habit formation. Habits arise through mimicking the action of others; hence their formation is the mechanism by which an individual ethos is constituted as a sign of the community ethos. Habit formation is a double-sided process. Through repetition of certain activities, corporeal affections (sensations) are internalised to become the property of the conscious self, and (conscious) feelings are given corporeal existence, or are physically represented, by shaping particular parts of the body and by enhancing bodily powers (ibid.: 140–3). So what is contingent (notably facial expressions and the voice) becomes a product of will, and what requires will, initially the will of the Other (e.g. standing) becomes mechanical. Habit is a 'being-at-home-with-oneself' (ibid.: 144). Habit formation is the means by which the distinctions between the inside and outside of the self, the self and body, the self and Other are constituted and it is also the process by which these interior and external surfaces of self are unified. Habit is the making of a subject who *is* his body.[4] Through habit, the individual relates himself to himself, removes himself from the contingency of sensation, and constitutes his corporeal individuality according to the will of all. The double movement of habit formation (the mental representation of sensations and the incorporation of feelings) is also how the body comes to represent mind, will, and thought. The body now moves and acts according to the dictates of the agent and the community – it is the sign of the will, both individual and universal. But not exactly.

Hegel sometimes tends to speak as if the body is simply moulded by an individual will which precedes it. The social will may precede and condition habits, but individual will does not. Non-conscious repetition of certain activities designated by others removes the difference between instances leading to a unity of bodily activity of which I am conscious as a unity separate from the world, others and myself (Hegel 1971: 144, 146). Unification of the self-possessed body through habit formation is therefore the precondition to the birth of *this* self, the ego, a self which is conscious of itself as a discrete entity and object for itself. Hence, individual will, agency and reflection open in the space between self and world, mind and body, self and other – a space constituted by the mechanism of habit formation. Individual will, as the practical mode of thought, is given its external expression (the habitual body) at the same time as it arises. Contrary to his own analysis, Hegel then claims that the body cannot adequately signify individual and universal thought because it is *natural* and particular (ibid.: 147). The body is hardly natural or particular given that it has been moulded to reflect the social will. What Hegel seems to push aside in an allegiance to universal subjectivity is his own insight: that individual will and thought are produced and maintained by a disjunction between mind and body, by the constitution of a difference between an inside and an outside of self. The habitual body cannot adequately signify thought (the will or ego) because the positing of the body's difference from self is precisely what gives rise to thought. And the habitual body cannot immediately reflect universal will because the body is what makes one an object or a sign for another and it is this positing of the body's difference from the Other (and the community will the Other represents) which gives rise to individual will.

Hegel's analysis of habits is about *how* the body is the site of one's specific ethos. As a sign of self, the body is always already socially constituted or transformed as it is invested with meaning and value. At the same time, because meaning is produced differentially, the body cannot stand alone as a sign of individual or universal will. There is an excess in the difference between mind and body and between the 'we' and the 'I' – a difference which arises from, but cannot be absorbed by, the relation between them. A difference to which reflection, language, and agency are living testimony. This reading of Hegel's theory of habit formation lends further explanatory force to Butler's claims, raised in chapter 2, that the subject can be said to be constituted but not determined by discourse and that the excess which accompanies the constitution of identity is the precondition of agency.

Discipline aims at the proper distribution of this excess (Hegel 1967: 117), and what is 'proper' for Hegel (and improper for Foucault) is the realisation of a will which is common to all and actualised in a unified social reality. The master/slave dialectic is the story of the actualisation of a unified social reality. It is also an extension of the story of how the identity of the self is constituted in and through another. It is the story of desire.

There are already indications in Hegel's account of habit formation of how the individual's identity is constituted in relation to others. As the individual, self-conscious self arises by constituting a difference between self and Other then that difference is contained within the self's identity and the self's identity lies outside of itself. The difference between mind and body, self and other, which I have located in the mechanism of habit formation as that which prevents the exact representation of the will (individual and universal) by the habitual body, is the cause of desire. In an attempt to overcome this difference, the subject, through desire manifest as language, work and action, continues to seek its identity in another: 'Self-consciousness is *Desire* in general' (Hegel 1977: 105). Or, as Judith Butler puts it: 'Human desire articulates the subject's relation to that which is *not* itself, that which is different, strange, novel, awaited, absent, lost. And the satisfaction of desire is the transformation of difference into identity' (1987: 9).[5] The transformation of difference into identity involves recognising the essential interrelatedness of the self, the other and the world and with this a transformation of the self, the other and the social reality which constitutes them. This transformation, told via the master/slave dialectic, is Hegel's version of how differences are accommodated through communal action. The question remains: is this accommodation possible and, if so, at what cost?

As the details of Hegel's 'master/slave' dialectic are already familiar, I will only highlight a few features of relevance to my discussion here. The most significant point, one usually overlooked, is that it is the body (as a sign of the self) which belies aspirations toward social unity based on the sameness of will. While the self posits itself as independent, it, like any sign, exists only in relation to another: 'Self-consciousness exists in and for itself when, and by the fact that, it exists for another; that is, it exists only in being acknowledged' (Hegel 1977: 111). But the other also claims an independent identity, signified by its specific embodiment, and will not passively provide the mirror-image the self seeks to project (ibid.: 112). So, in order for the self to be certain of its own self-representation in the other, a double action is required such that both can '*recognize* themselves as *mutually recognising* one another' (ibid.: 112). This mutual recognition, this self-understanding in the face of a transparent other, requires the establishing of equivalences. The other must become what I value in myself, my self-image, and I must become what the other desires (Kojève 1980: 7). But each appears to the other as a body which signifies a difference from the other and dependence on the contingencies of life (Hegel 1977: 113). The body must disappear as an essential element of one's ethos if the self is to succeed in projecting and reclaiming a representation of itself as independent and universal. The body must be negated because it hinders the establishment of equivalences which can be mutually exchanged through recognition. And, to reiterate the point established earlier, what is negated with the body is not a natural sign but a socially constituted mode of being which carries habits, capacities and interests specific to one's ethos.

To prevent any misunderstanding, then, to ensure self-transparency, the self must represent itself to the other through an action which shows 'that it is not attached to any specific *existence*' (Hegel 1977: 113). In contemporary terms this action would involve acknowledging the interrelatedness of social identity and abstracting oneself from one's specific embodied ethos in an attempt to have one's position, and that of the other, recognised. Hegel's method of abstraction is a little more dramatic. The double action required for mutual recognition (and the constitution of the 'I' that is 'we') is that each must seek the death of the other (which affirms the identity of each as essential) and each must stake his own life (which confirms that the difference signified by one's embodiment is inessential). However, while the body may be an inappropriate expression of the sameness of identity, it cannot disappear. One's self-representation can only be confirmed through the body as a sign of the self and in the face of another who lives to recognise it (ibid.: 115).

This struggle for mutual recognition is not the battle between the sexes but a battle between two equal self-consciousnesses, both of whom have negated the significance of their embodiment – they have negated their differences. But mutual recognition does not flow immediately from this negation. With the necessary preservation of embodiment, the desire to kill the other is replaced by a desire to dominate the other. As Judith Butler suggests, in this relation of domination one party (the

'Lord') comes to signify universal, independent subjectivity, and in claiming independent identity must project on to the other (the 'Bondsman') all that the Lord 'endeavours *not* to be': the appearance of being 'an unfree body, a lifeless instrument' (Butler 1987: 52–3). As embodiment is what seems to prohibit mutual recognition, one would assume that the body is what is transformed in order to reflect and represent the 'I' – that is, universal subjectivity or what Hegel calls the 'Lord'. But this is not the case, at least not simply. It is through *work* (by the body/Bondsman) in the service of the presence of unified universal subjectivity that 'desire is held in check', displaced, and the 'outside' of materiality is transformed in order to reflect the self (also transformed in the process). The making-over of the material world into a sign of the self is how social reality is transformed to represent the will of all, and this is the precondition to mutual recognition, to the exchange of equivalences, to ethical life.

Yet, within this theatre of social production, the body is not merely a prop. For Hegel's analysis of the work of desire to be complete, one would assume that it is not just the subject's external environment and mode of consciousness which are transformed such that they reflect each other. The precondition of mutual recognition would also involve the reshaping of the subject's own embodiment so that the self is for-another as he is for-himself. If there is to be a union of the will and the world, the body, as the immediate sign of self, must also be refigured. Hegel suggests as much in his additions to the *Philosophy of Mind*:

> Body is the middle term by which I come together with the external world as such. Consequently, if I want to realize my aims, I must make my body capable of carrying over this subjectivity into the external world.
>
> (Hegel 1971: 146)

Habits, capacities, and bodily powers must be rewritten in the translation of universal subjectivity into social reality. While communal action, or the work of desire, may reconstitute sociality to accommodate differences, the differences so accommodated are limited: Hegel's social unity is not only based on the assumption of the universal subject but also on the sameness of bodies. In particular, as I will go on to argue, in Hegel's ethical life the body of the community thus transformed does not represent or directly reconstitute the body – nor, therefore, the will – of woman.

While the male ethos is constituted in this process of social transformation, there is still a difficulty in claiming that sociality can adequately represent the male body and the individual will it signifies. The assumption of sameness, while informing social relations and the constitution of the individual, is not fully actualised. As I have already argued, the difference between mind and body, self and Other generated in habit formation is the precondition to will, ego and thought. And the other's difference or independence, signified by the other's body, is a precondition to the desire for self-representation. Hence, if reflection and its symptoms (language, desire and action) are maintained then so is this alterity. Hegel's reduction of individual action and social reality to signs of 'rational' thought cannot account for an alterity which resists as it motivates the unity of identity and difference. The male subject remains a duality which fails to reach unity with itself – a point not lost on Hegel.[6] However, as I will argue in the following analysis of Hegel's 'ethical life', the pretence of individual autonomy within a social unity is upheld by excluding from the community those bodies, parts of bodies, or bodily activities which cannot represent the sameness of will. And this dependence of social unity upon the social representation of one embodied ethos has certain effects on the constitution of the bodies of women.

HEGEL'S 'DIALOGICAL' ETHICS

Hegel's thesis that the self's identity is constituted through the Other allows an account of how an embodied ethos is constituted to reflect that of the community and vice versa. I have suggested, on the one hand, that as the self comes into being through the Other, it is always other than itself. On the other hand, Hegel wants this otherness overcome: he hopes for the unity of identity and difference where the work of both habit formation and desire would effect the normalisation of bodies and the transformation of social institutions such that the whole of social reality comes to reflect the self. The first moment of this unity is ethical life. But there is an internal difference within ethical life: sexual difference. In his account of ethics and sexual difference Hegel makes explicit what is implied in his analysis of desire: that women are excluded from the exchange of equivalences necessary for social unity, although not from their construction.

What I will argue below is that this exclusion of women is not arbitrary: the social unity Hegel posits is predicated upon this exclusion. Insofar as the thesis of a social unity assumes an individual identity which is transparent to itself and to the Other (the unity of identity and difference) it assumes differences which are really variations of the same. In assuming that the other's difference is transparent to and contained within that identity which is the ethos of the community, the value of this ethos is maintained by the exclusion of others. That others (women in Hegel's case) are excluded suggests, not the existence of differences which, as accidents of nature, don't measure up to the norm, but differences constituted as a remainder to the normalising process.

Having said that women are excluded from the exchange of equivalences by mutual recognition, I should add that, paradoxically, they are not excluded from ethical life. In fact Hegel claims, using the story of Antigone, that ethical life is predicated on the unity of sexual difference. To begin to make sense of this paradox and to illustrate what is at stake in any ethics which moves towards the unity of identity and difference, let me reiterate the story of Antigone.

The drama of Sophocles' play opens with a conflict which Hegel ignores in his rendition of the story. The conflict is between two sisters: Antigone and Ismene. Antigone, in defiance of a decree issued by her uncle (Creon), the ruler of the state, proposes to bury her brother, Polyneices, who has been killed in a battle against the state. Ismene begs Antigone not to sacrifice herself in enacting her ethical duty toward her brother. The contemporary actors in this play would be: the woman (Antigone) who, in expressing her right to moral autonomy within patriarchal society, implicitly upholds the individual male ethos as the universal; and the woman (Ismene) who claims that femininity signifies a difference which is more than the other side of man.

Antigone, the archetypal feminist, appears to be asserting her independent will against the laws of a patriarch. Her reward for defiance of the law will be, she assumes, equal status with her brother (Sophocles 1960: 63). But such self-expression does not go unpunished: the penalty for burying Polyneices is to be death by public stoning. In contrast to Antigone, Ismene appears to be little more than a passive fool who accepts her subordinate position within a male dominated community: 'since I am no free agent, I will yield to the powers that be' (ibid.). But appearances can be deceiving. While Ismene advises Antigone not to take on the state, she does so from concern for her sister's fate and from a sharper appreciation of what is at stake. All that Antigone's proposed action will evoke is a universal 'ethical' law which bestows honour upon the male body. For Antigone to sacrifice herself to such a cause is hopeless, according to Ismene (ibid.: 64). She suggests that, if Antigone must act against the state, she should do so secretly in order to preserve herself (ibid.).

The dispute between Ismene and Antigone is parallel to an apparent impasse facing contemporary feminism. Demands for equality and autonomous self-representation may have improved women's lot, but at the expense of further entrenching an implicit male norm to women's disadvantage. There is also a suggestion of a residue in the operation of sexual 'difference' which is not subsumed by this norm. But, if we follow Ismene's advice, this difference can only be preserved through secrecy and silence – hardly a satisfying 'solution'.

The argument between Ismene and Antigone is surpassed in the play (only to return after Antigone's deed is done) by the conflict between

Antigone and her uncle. Creon seeks to preserve the unity of the community against any expression of individual difference (in this case, that of Polyneices), while Antigone implicitly destroys that unity by building a shrine in honour of Polyneices' (individual male) body. This battle between the sexes becomes the centrepiece of the tragedy, resulting in the downfall of both.

For Hegel, *Antigone* is a picture of ethical life in which individual action is a moment of difference within a self-restoring unity (Hegel 1977: 260–1). Hegel can only sustain his reading by pushing the conflict between Ismene and Antigone offstage. He makes the tragedy of sexual difference conditional upon another relation: that between twin brothers, Eteocles and Polyneices (ibid.: 285–6). The ethical significance of this relation is that a 'natural' difference has appeared which must be unified (consciousness is split accidentally between two identical but different bodies). But the unity of identity and difference does not occur without a fight. The opposition between Eteocles, who represents the unified community, and Polyneices, who signifies individuality against the state, leads to the death of both. The conflict between Creon and Antigone is over what to do with Polyneices' body, a body which has challenged the unity of the community. So, as I will go on to argue in more detail, sexual difference is understood in terms of a male ethos: femininity is said to be the ethos which upholds male individuality; masculinity is a sign of the communal ethos.

Hegel hopes for a reconciliation between the sexes and therefore between the male individuals and the community their positions are said to represent. He hopes that through the conflicting actions of Antigone and Creon their differences will be dissolved and society will be transformed toward a higher unity. This reconciliation depends on a number of assumptions I have already questioned. First it assumes that action is a transparent sign of individual will which, in turn, is a practical expression of thought (Hegel 1977: 281). Action for Hegel, no less than speech, desire and work, is a mode of self-representation. Action signifies one's ethos, not because the will comes before its signifier but because, through action, the self is both constituted and expressed: 'the deed is the *actual self*' as Hegel puts it (ibid.: 279). While accepting the claim that the self does not come before its social signifier (action, language, etc.), I question the claim that the self is transparent to others through action. This claim that the sexes can mutually recognise each other through action is based on a second, related, assumption: that the individual's ethos (including that of men and women) is contained within and therefore transparent to that of others and that differences are partial signs of, and are expressed within, the communal ethos. Ethical social reality is such that 'Ethical *self*-consciousness is *immediately* one with essential being through the *universality* of its *self*' (ibid.: 261). This universality of self is a society produced through the action of all and where the individual is a moment of this action (ibid.: 264–5). So, when action represents an 'individual that is a world' (ibid.) life is ethical because the individual enjoys moral autonomy within a social and political structure which is no longer alien to their action. The actions of men and women, for example, can be in opposition, but if they make any sense at all they are supposedly transparent to each other and accommodated within the communal ethos.

A contemporary ethics of difference, which claims that we can dissolve or accommodate our differences by expressing them, would approve of Hegel's reading of *Antigone* without the loss of blood. John McCumber (1988), for example, champions Hegel's critique of ethics based on a pre-social will but objects to the normalising tendencies in Hegel's own ethics. He argues that as moral agency on Hegel's model is not grounded in individual will but in a unified social tradition of which the individual is a product, then the individual is reduced to a passive node in a matrix of social institutions and practices – a view he also attributes to Foucault (McCumber 1988: 133). As individual experience and action is already dictated by the language and practices a society embodies (*Sittlichkeit*), then, in Hegel's ethics, 'unique experience' is safely walled away in the realm of the ineffable. For McCumber, 'unique experience' is the site of difference and its expression is a precursor to social change (ibid.: 142–3). However, it is not clear where McCumber thinks 'unique experience' comes from, given that like Hegel he rejects the assumption of a pre-social will. He does suggest, in apparent opposition to Hegel, that we adhere to a multiplicity of *Sittlichkeit* rather than a single all-embracing ethical unity. This, however, is his model for how 'unique experience' can be expressed, rather than its origin. In any case, according to McCumber, when faced with competing ethical possibilities (the multiplicity of *Sittlichkeit*) we make ad hoc modifications to language in order to articulate our unique experiences to others. This ad hoc 'dialogue' allows the representation of differences, is the site of moral agency, and is necessarily disruptive. McCumber's 'dialogical' ethics, based on 'reflective reflection' as he calls it, is a formula for a 'counter-practice' to the social norms which embody the status quo.

A dialogue, as a means of representing differences, is a practice not restricted to McCumber's design for a post-Hegelian ethics. As already mentioned in chapter 1, some feminists who object to 'universalist' ethics also propose dialogue as a solution to the exclusion of the representation of sexual difference from social discourse. Besides an ethics based on 'interactive rationality', suggested by Benhabib, there is also Susan Parsons' proposal for a feminist ethics which promotes an 'active dialogue between the changing aspects of our humanness amidst various social interactions' (Parsons 1987: 11). By this I take her to mean that, instead of adhering to universal moral principles which do not allow for social change and which leave differences out of account, we should remain sensitive to the specificity and changeability of our own social context as well as that of the 'others with whom we share life'. Such a 'dialogue' deepens 'one's insights into present social realities', promotes 'self-understanding' and generates 'new meanings' and different directions for the future (ibid.: 10–11). Jana Sawicki, with some reference to the work of Foucault, suggests that in 'a feminist politics of difference, theory and moral judgements would be geared to specific contexts' (Sawicki 1986: 35). That is, rather than subsuming all differences under social norms, we need to redefine our own differences and 'discover distortions in our understandings of each other and the world' (ibid.: 32). This mutual understanding is reached through dialogue (Sawicki 1988: 187). A dialogue can be opened between different women, for example, not with the aim of eliminating differences but of better understanding the differences and similarities between women and 'discovering the basis of coalition building' (ibid.).

While sympathetic with these attempts to evoke an ethics of difference, I am not convinced that the notion of dialogue between the privileged and the 'ineffable', or between particular individuals, overcomes the normalisation apparent in Hegel's ethics. While Hegel's ethics ignores differences between women it shares other assumptions with a dialogical ethics. Both would allow that Antigone and Creon, for example, are constituted differently as a result of being located in different positions within social discourses and practices. Both would also allow that through dialogue or joint action the differences between the sexes could become transparent to each other, understood and accommodated with attendant changes to social and political institutions.

There is no doubt that Hegel's ethics is normalising. But this is not because he leaves 'unique experience' or difference safely walled away in the realm of the ineffable, as McCumber suggests. Rather it is because of what he claims happens to this difference in its expression. As I have argued in my reading of Hegel on habits and desire, insofar as he allows that individual will is an effect (rather than cause) of the self being distinguished from the other (other people and the world) then there is agency and difference. And no sign of the self (action or words) can adequately express the specificity of self. Not because 'unique experience' comes before a language or action which effaces it, but because one's specificity is neither fully present to the other nor simply ineffable.

In acting, for example, the self is not just represented, it is actualised. Yet, while constituted in the act, the self is not produced with a self-contained meaning or identity. Within the act, the self is divided (Hegel

1977: 266) because the identity of an ethos, like any sign, is determined by its relation to what it is not. The specificity of one's ethos is realised as pathos (ibid.: 287). One's position is a dis-position, in that the self, expressed in action, evokes its debt to the other and accrues guilt through a crime against the other (ibid.: 281–2). On the one hand then, Hegel's analysis of action suggests that in expressing the specificity of self, the self is further dispersed in another. All action or speech is therefore to some extent ineffable in that one cannot be at the same time the self expressed in action and the reader of that text.

On the other hand, Hegel does assume that, through action in opposition to others, we become aware of ourselves in that which we oppose (Hegel 1977: 284) such that the particular self is dissolved in action into a destiny, the truth of which is the universal self (ibid.: 285). This normalising of differences rests on the assumption that the self, and the other, becomes transparent to itself in action and speech. Insofar as a 'dialogical ethics' or an ethics of communal action also assumes self-transparency and transparency of the other in the accommodation of differences and/or pre-social agency, then it shares, with Hegel, this tendency towards universality and normalisation. A dialogue which claims absolute understanding of the other is, in effect, a monologue which subsumes differences under norms already in place. The social fabric may alter as an effect of dialogue and action but the inequalities within it will remain in place. And this is nowhere more apparent than in the case of sexual difference.

The following analysis of the fate of sexual difference in Hegel's ethics is not an argument against attempts to accommodate differences through dialogue with the excluded or through community action. Rather it is an argument against the assumption that differences will be automatically represented in this way without loss of meaning. It is an argument for the need also to interrogate the conditions under which the privileged position of some men may be maintained precisely through dialogue and communal action.

THE IRONY OF WOMEN'S BODIES

While, for Hegel, 'every individual may be a child of his time' (Hegel 1967: 11) not everyone enjoys 'the habit of right and goodness [as] an embodiment of liberty' (Hegel 1971: 192), at least not in the same way. While the habitual body is constituted within, and as a reflection of, a social structure, not all bodies have the same shape or develop the same habits and capacities. In particular, the different habits of men and women reflect their different positions within the social text. Hegel's own analysis of ethics reveals that women's bodies are the ground for the material construction and exchange of equivalences between men.

But there is something about the habitual body, socially constituted as signifying 'woman', which accords her a value inappropriate for this exchange. Yet women are not simply excluded from ethical life or confined to the realm of the ineffable. The exclusion of women's bodies from equitable social representation allows the assumption of social unity. At the same time, as I will go on to argue, women's differences thus constituted undermine that possibility of unity: 'woman' is the irony of the community.

Hegel makes the astute observation that women, as wives and mothers, are deprived of social representation for, and therefore recognition of, their independent subjectivity. There is (supposedly) mutual recognition in the relation between husband and wife but this is a 'natural not an ethical one' (Hegel 1977: 273). It is not sustained as socially significant; it does not register in the communal ethos. Rather, the husband/wife relation has a meaning which gains its existence in something other than itself – the child. Insofar as a woman does achieve any social expression it is via the independent social activity of her child or her husband. Her desire for independent self-representation is subordinated to her ethical duty of supporting the activities of her husband and children insofar as they maintain the universal (community) (ibid.: 274–5).

Men, on the other hand, do not have to sacrifice their desire. Their individual self-representation is catered for in private by women, and their work in the service of the communal ethos, and hence their universal self-expression, is undertaken in the public sphere. This suggests not only that the universal will represented by the community is exclusively male, but also that it is maintained by excluding and devaluing others.[7]

That a female ethos is not represented in public institutions is partly an accident of history – traditionally women have been excluded from citizenship. But, according to Hegel, this exclusion is also structurally necessary. The communal ethos, like any sign, does not exist in-itself – its meaning and value is derived from its relation to what it is not. Contrary to McCumber's claim that Hegel posits a simple all-encompassing ethical unity, *Sittlichkeit* is internally divided between a social reality and its 'others'. A community is an individual entity, for example, in relation to other communities (Hegel 1977: 288) so that a truly universal community is only realised through war which dissolves the difference. And, more important for the question at hand, the ethos of one community is divided between what Hegel calls Human Law and its unconscious Notion, Divine Law. Human Law is individual (male) self-consciousness realised as a social body in which the citizen is conscious of his actualisation in its customs and laws (ibid.: 276–8). In other words, the community is the 'pyramid' or the body of the sign

– a body which has been transformed through the work of (almost) all to represent the universality of self. The soul of the sign, its meaning, is male consciousness. But the body of the sign is also the individual male body, the 'tip of the pyramid', which resides within the family, and its soul is, oddly enough, female consciousness (Divine Law). The male ethos needs its other to remain 'other' for its own privilege to be maintained.

While not immediately obvious, the principle which links the family and the community is the same as that which relates the individual to the communal ethos: the negation of bodily difference. The stability of the community relies on maintaining a separate sphere (the family) where women cater for the male individual's particular bodily needs and desires – those not appropriate to the unity of the community. The family also provides the community with citizens, and for this the community has a relation to the family which 'consists in expelling the individual from the Family, subduing the natural aspect and separateness of his existence, and training him to be virtuous to a life in and for the universal [community]' (Hegel 1977: 269). As discussed, this training consists in the formation of habits, a moulding of the male body into a sign of community will. The negation of bodily difference also involves the work of death. The (potential) death of the body, as Hegel indicates in his master/slave dialectic, carries the significance of elevating the individual above the contingency and particularity of 'natural' existence. The work of death universalises the male body. In the community this is realised not just through habit formation but also through war, where the individual male body is staked and negated as inessential.[8] Similarly, the ethical duty of women in the family is to raise the male body to the status of the universal. This involves the care of the male body in general and includes adding the significance of universality to natural death through the public administration of death rites (ibid.: 271–2). Through the work of women in the family, the male body as the site of 'difference' is transformed into the sign of the universal without disrupting a unified social reality.

The body of the sign, then, is split between the body of the community and the individual male body. Its meaning is divided between two laws represented by male and female consciousness insofar as they are ethical (Hegel 1977: 275). The difference between the two laws and hence the two sexes is, in Hegel's model, one of complementarity where 'woman' is reduced to man's other. True universality or social harmony resides in the unity of this difference:

> The difference in the physical characteristics of the two sexes has a rational basis and consequently acquires an intellectual and ethical significance. This significance is determined by the difference into

> which the ethical substantiality, as a concept, internally sunders itself in order that its vitality may become a concrete unity consequent upon this difference.
>
> (Hegel 1967: 114)[9]

Ethical unity is predicated on a sexual difference which is bodily and which signifies a difference in duty or will. Enacting this duty, according to Hegel, dissolves, and thereby accommodates, the difference.

As I have mentioned, the self is not actualised within the community ethos without action. Using the story of Antigone, Hegel illustrates how, in actualising one law through action, each sex becomes aware of its opposition to that of which it was ignorant – the other law. Through action the other becomes transparent to the self, as does the essential relation between the two. By evoking the work of death through war, Creon (who embodies the state) 'knocks off the very tip of the pyramid' (Hegel 1977: 286). That is, the community negates the significance of the individual body (in this case, that of Polyneices) which threatens its unity. Antigone, in defiance of the state, but in keeping with her ethos, restores her brother's body to the status of the universal through death rites. The actions of the two sexes signal the demise of each. But, as the two laws are united in essence, the dissolution of sexual difference and the laws they embody results in a new unity – a universal self which absorbs both. And, through antagonistic actions, the community is transformed into a new ethos which has absorbed the different positions.

In effect, the difference dissolved here is that between male individuality – upheld by women – and the body of the community. This, according to Hegel, is all there is to sexual difference. On his model of identity and difference the interval or difference between the sexes is constituted such that man's identity is upheld and unified by woman understood as his complement. But the success of this dialectic depends on whether Hegel's understanding of sexual difference as complementarity is adequate. This hinges on whether, on his own analysis, the ethos of each sex is the immediate sign of an opposition which is part of, and absorbed by, the whole. And this depends on the degree to which women's bodies signify a duty to a body not their own.

Ethical action, including Antigone's, is, for Hegel, an expression of individual will which is already represented in social reality. Yet women, as wives and mothers, are, according to Hegel's own account, denied such self-representation in order that a male ethos be maintained as the norm. According to Hegel, women are constituted through the suppression of individuality:

> Since the community only gets an existence through its interference with the happiness of the Family, and by dissolving (individual) self-consciousness into the universal, it creates for itself in what it

> suppresses and what is at the same time essential to it an internal enemy – womankind in general. Womankind – the everlasting irony (in the life) of the community – changes by intrigue the universal end of the government into a private end, transforms its universal activity into a work of some particular individual, and perverts the universal property of the state into a possession and ornament for the Family. (Hegel 1977: 288)

What is suppressed by the community in the first instance is *male* individuality. Women are the enemy of the community insofar as male individuality becomes their end. But if femininity is coextensive with the promotion of male individuality, then this suggests that woman is constituted by default – through the suppression of the representation of her own difference. It is not so much male individuality that is a threat to the unity of identity and difference, for the community can dissolve this difference through discipline, war and other manifestations of community spirit such as dialogue or action. Nor is femininity an everlasting threat if it is coextensive with raising the status of the male body to the universal. Even though such action is a threat to communal unity in the short term, the universal will be realised in the end. But the universal Hegel has in mind is a more adequate representation of male consciousness – more adequate than its immediate representation in the male body. Hence the eternal enemy of Hegel's community is the representation of a difference which cannot be absorbed into the universal male ethos he assumes.

How then is this female ethos constituted so that women embody 'Divine law' and will automatically act to ensure the universal representation of a body other than their own? And, following from this, is there another aspect of sexual difference irreducible to that which upholds a dominant male ethos? According to Hegel, a woman's ethical action is grounded in her constitution of self in relation to her brother: a relation to a man based on mutual recognition rather than sexual desire, a relation where the woman is no longer a daughter and not yet a wife.

> They are the same blood which has, however, in them reached a state of rest and equilibrium. Therefore, they do not desire one another, nor have they given to or received from one another this independent being-for-self; on the contrary, they are free individualities in regard to each other. . . . The brother, however, [unlike the husband] is for the sister a passive, similar being in general; the recognition of herself in him is pure and unmixed with any natural desire. In this relationship, therefore, the indifference of the particularity, and the ethical contingency of the latter, are not present; but the moment of the individual self, recognizing and being recognized, can here assert its right, because it is linked to the equilibrium of the blood and is a

> relation devoid of desire. The loss of the brother is therefore irreparable to the sister and her duty towards him is the highest.
>
> (Hegel 1977: 274–5)

Hegel counts on an ideal constitution of sexual difference – a difference which is not really a difference given that a woman, as a result of this relation, will be devoted to the universalisation of a body as a sign not of herself, but of her brother. This original harmony of mutual recognition is not exactly original, for the two moments come together already constituted ('they are free individualities in regard to each other'). Nor is it exactly mutual. As Luce Irigaray suggests, the brother-sister relation

> is a consoling fancy, a truce in the struggle between uneven foes, a denial of guilt already weighing heavily on the subject; . . . both sexes, male and female, have already yielded to a destiny that is different for each . . . as Hegel admits when he affirms that the brother is for the sister the possibility of recognition of which she is deprived as mother and wife, but does not state that the situation is reciprocal. This means that the brother has already been invested with a value for the sister that she cannot offer in return, except by devoting herself to his cult after death.
>
> (Irigaray 1985: 217)

Irigaray is suggesting that there is a difference already operating in the brother–sister relation which is not absorbed in that relation nor in the duty which proceeds from it. Not only does the exclusion of women's bodies from public life maintain the pretence of unity and male autonomy within the community, but there seems to be in operation a difference which exceeds woman's different habit of will upon which complementary difference counts and which is suppressed and denied in the constitution of the interval between man and his complement. What, then, is the source of this other difference?

It is not at all clear how Antigone obtained her 'independent being-for-self' so that her relation with her brother is one of mutual recognition. Hence it is not clear that her duty, as a result of that relation, will be exclusively to his body. From Hegel's discussion of habits we know how the will arises, how the body comes to (almost) represent the will or ego and how the drive for exact and substantial self-representation generates desire and action. We also know that women are denied direct access to much of the social means of representation open to men. Hegel says little more than this in the *Phenomenology* about women's path to a different individuality. However, in the *Philosophy of Right* he claims that 'the difference between men and women is like that between animals and plants' (Hegel 1967: 263). And he describes a plant elsewhere as follows:

> Even in the plant we see a centre, which has overflowed into the periphery, a concentration of the differences, a self-development from within outwards, a unity which differentiates itself and from its differentiation produces itself in the bud, something, therefore, to which we attribute an urge (*Trieb*); but this unity remains incomplete because the plant's process of articulating itself is a coming-forth-from-self of the vegetable subject, each part is the whole plant, a repetition of it, and consequently the organs are not held in complete subjection to the unity of the subject.
>
> (Hegel 1971: 9–10)[10]

This description of a difference proceeding from a unity and then aspiring to a higher unity through repetition is reminiscent of Hegel's theory of habit formation where the 'sentient self' is transformed into a 'self-centred subject'. Women then embody a differentiation which allows a 'rapport' with, and subjection by, others and a moulding of the body to reflect the self. There is also an 'overflowing into the periphery', an excess of self-development, which is the will. But apparently a woman is a vegetable and does not achieve the unity between the body and the self, the outside and the inside, which supposedly occurs in habit formation.

The possibility of men attaining this unity of self is at best uncertain, as I have suggested. For women this uncertainty is doubled. They may undergo some discipline and education, but 'who knows how' says Hegel (Hegel 1967: 264). In order to sustain a male ethos as the norm and thus give men a chance at self-unity, women are denied those means of objectification of will open to men.

> Thus one sex [man] is mind in its self-redemption into explicit personal self-subsistence and the knowledge and volition of free universality, i.e. the self-consciousness of conceptual thought and the volition of the objective final end. The other sex [woman] is mind maintaining itself in unity as knowledge and volition of the substantive, but knowledge and volition in the form of concrete individuality and feeling. In relation to externality, the former is powerful and active, the latter passive and subjective.
>
> (ibid.: 114)

The ethos open to women in Hegel's ethical life involves the care of particular others. So, as wives and mothers, women may develop habits and patterns of existence, but these are disrupted and subordinated to the habits of others. Thus, the 'rapport' apparent in the sentient self's relations with others, which is internalised in men through habit formation (as a self-relation) is not internalised to the same extent in women. (Hegel suggests, for example, that this rapport manifests in conscious life particularly between female friends (Hegel 1971: 95).) This dispersion of self into the other may account for an ethos of 'interrelatedness' attributed to women and valorised by some feminist ethics (e.g. Gilligan 1982). But in a man's world the excess of self or will created through differentiation from others remains suspended in women (woman is 'mind maintaining itself' in the 'form of feeling'). The will is turned in on itself and the body, moulded according to repetition of this excess, remains a woman's only means of self-representation. In Hegel's ethical life a woman embodies what is other to the body politic and her body repeats and intensifies its difference in contrast to the will represented by the body of the community.

Simone de Beauvoir (1972) agrees that this is the effect on the constitution of woman's body of maintaining a male ethos as the norm. For de Beauvoir, 'woman, like man, *is* her body' (de Beauvoir 1972: 61). However, the cultural production of woman's body is not a happy one. Men and women are positioned differently: man is the Subject and woman his Other (ibid.: 16). Men are in a position to transcend their present embodiment and create values through the work of death whereas women are destined for the 'repetition of Life' (ibid.: 96). Woman may be her body, but without the means of projecting herself, transcending herself and distinguishing herself from the Other, 'her body is something other than herself' (ibid.: 61). As man's other, woman, unlike man, is not at home in her embodied ethos.

If we accept for a moment that, in the interests of maintaining a male ethos as a norm, a woman's ethos is constituted as dispersed into others and other to herself, there are two points to be made. First, how should we proceed to undo the source of women's alienation from themselves? De Beauvoir suggests that women too can be 'at home with themselves' if they rise above immersion in the body and in the other and attend to the same worldly projects as men (de Beauvoir 1972: 725–41).[11] Joanna Hodge in her reading of Hegel suggests a similar solution: the tensions generated by Hegel's rationalisation of the exclusion of women from political and cultural life can be removed by extending 'the consciousness of freedom to women' (Hodge 1987: 129).[12] Or there is the possibility of a dialogue between the sexes by which their differences could be represented.

This would seem to be Antigone's project: to assert herself in social life by expressing her difference from men. If this difference is understood then it is based on a relation of mutual recognition (initially with her brother, if we follow Hegel). But, and this is the second point, if Antigone is in the state of suspended animation that is a woman's ethos, when she enters into a relation with her brother, if her body is the repetition of a difference from a male ethos, then it would be surprising if she found any recognition of herself there. His body, as the immediate sign of his will, has undergone a different mode of production. On the other hand, if there is a dialogue between the two where mutual recognition and understanding is achieved, then Antigone's difference will be transformed and subsumed under norms already in place. Given that the communal ethos is based on the reproduction of a male body, then in the absence of any alternative mode of externalisation, Antigone's self-expression will involve care of her brother's body and the universalisation of masculinity this implies.

Paradoxically, however, within this ethos of care, woman does not exist: as Creon remarks in the wake of Antigone's action 'she is a man' (Sophocles 1960: 77), and Antigone herself admits that she has been 'dead for a long time now' (ibid.: 79). So where are women really? Contrary to Hegel's suggestion, women are not vegetables, pregnant with a meaning which awaits an action to realise man's self-image. If this was the case then Hegel would be right in suggesting that woman's action, as the other side of man's presence, would result in a dissolution of difference into a whole. If difference could be reduced to complementarity then a 'dialogical ethics' would be right in assuming that differences could be accommodated through mutual self-understanding. But the divisions within the social structure which uphold male subjectivity as the universal and leave women without habitually produced borders of self also encourage women to seek self-expression in projects other than those directly related to men. Hence, it is not just man as other to, and therefore part of, the self that women recognise in the result of their action.

In Antigone's action there is a remainder not absorbed by the universal. Her body does not mirror her brother's, nor is her action exhausted by her duty in the service of masculine individuality. Her body remains after her action and cannot be universalised through integration into the community or through the private work of death ('Antigone', as the Chorus notes, is 'an alien in both worlds' (Sophocles 1960: 89)). There is no public punishment to fit her crime: rather than carrying out the penalty of public stoning, Creon shuts 'her in a rocky vault alive . . . to stop pollution falling on the city' (ibid.: 86). Nor is Antigone's will confined to her action in the service of her brother. She goes on to defy the structure which produces, then disavows, her embodied ethos by taking her own life. She thus contaminates the body politic with her body as a sign of herself and deprives the state of its means of reproducing itself through its sons (Antigone was to marry Haemon, Creon's son, but he takes his own life as a result of her death). Hence, woman is, like man, other than herself, but the sign of her difference cannot be simply absorbed in a path to social unity. Nor is her meaning ineffable in contrast to the full presence of man's meaning. Rather, her difference is necessary to and a product of – but at the same time defies – a social 'unity' which represents a body other than her own.

There is another mode of externalisation open to women, other than the brother–sister, husband–wife, or mother–child relation, which Hegel symptomatically omits from his dialectic. This is the mode of self-representation predicated on the relation between sisters. Ismene's 'overflowing' of 'self-development', for example, has been displaced outside herself. This displacement is divided between duty to the state (Sophocles 1960: 63) and allegiance to her sister, Antigone. Above all it is in Antigone, rather than in her brother, that Ismene recognises herself, so that her sister's action/crime becomes her own. 'If my sister will allow me', says Ismene, 'half the blame is mine'. But Antigone will have no part of this recognition (Sophocles 1960: 78–9). She rejects Ismene's claim on her 'independent' selfhood and the action she partially embodies. This antagonism is not resolved by the battle between the sexes. Ismene's body remains suspended between that which reproduces masculinity and that which remains other. And insofar as Antigone's action is not absorbed by the dissolution of the opposition between the sexes, then neither is Ismene's mode of self-representation.

Antigone and Creon are eternal figures insofar as our body politic still represents and is represented by a particular kind of male body. What I have attempted to show, using Hegel's account of the production of identity and difference, is how this mode of representation operates and the nature of its material effects. The reproduction of a male ethos as the norm still depends on the exclusion of others. One doesn't have to go far for examples of how female modes of embodiment still cannot easily participate in the body politic without disruption. Pregnant bodies, menstrual bodies, and menopausal bodies have yet to find a comfortable public place.

It could be argued, however, that, notwithstanding these exclusions, women now enjoy more independence within public life. We do reconstitute our bodies through a change of habits – gymnasiums abound with examples of this possibility, as do disciplinary structures within the workplace. But if Antigone symbolises the possibility of 'extending consciousness of freedom to women', as Hodge implies (Hodge 1987: 152), if she is de Beauvoir's 'her' (the 'I' which seeks social representation through communal projects), then her body is still other than herself. We 'enlightened' women of the 1990s may no longer describe women as Hegel or de Beauvoir do or seem to live our bodies with the shame they suggest. But if it is still the case that a male embodied ethos is the norm then we have yet to break the habit of sexual 'difference' which constitutes women's bodies as not-yet-at-home, as alien to woman's best interests and to those of public life. Nor does valorising women's apparent shared moral perspective of care and interrelatedness overcome these difficulties: such an ethos is based on the same process of alienation – on women's representation of, and duty to, a body other than our own.

This leaves us with the ethos of sisterhood and the possibility of accommodating differences between women through dialogue. If Antigone and Creon are eternal figures, then so is Ismene. The relation between Ismene and Antigone may be based on the apparent sameness of a different body, but it does not satisfy the desires of either. The relation is socially constituted and therefore involves some complicity in upholding masculinity – allegiances are split. But even if this is denied, even if we accept the claim that there is a femininity outside the complementary difference of patriarchy, there is no unified 'I' or 'we' that precedes or results from relations between women. A feminist critique of complementary difference, whether directed at Hegel or at the tradition of thought which follows him (including psychoanalysis), must also come to terms with the consequences of this critique as it applies to women. A woman's specific ethos depends, no less than a man's, on the constitution of a body as a sign in relation to others such that its meaning is dispersed at the moment it comes into being. Hence, even a woman's mode of self-representation in relation to other women is steeped in misunderstanding, and points of commonality between women are necessarily temporary. To claim otherwise is to further efface differences between women. The tendency to assume an ethos of being-in-the-world-among-women, to assume that women shouldn't or don't objectify other women or that differences between women can be understood through dialogue is the crime we commit against our sex without necessarily knowing it.

The fact that women's bodies defy the condition of a 'total womanhood' which would universalise a male ethos indicates how the process by which the body comes to represent the self undercuts Hegel's hopes for the unity of identity and difference. The social production of bodies disrupts totalisation. At the same time, Hegel's assumption that difference is the other side of sameness leaves women in a difficult position within his ethics.

In the following chapter I will explore attempts to open up Hegel's limited economy of difference to a consideration of differences beyond mere opposition and complementarity. These differences, as my discussion of the relation between Antigone and Ismene indicates, do not lie outside the social production of a dominant male ethos. They are produced within this economy, yet exceed its terms.

Chapter 4

Sexual difference beyond duality

In Hegel's model of the constitution of identity and difference sexual difference is understood as opposition, complementarity or duality. A male embodied ethos operates as the norm of social identity and women are constituted as man's 'other'. There are two points about this model of sexual difference which bear repeating: (1) man's self-identity, the meaning and value of his ethos, is given to him through the work of women who universalise his body and remain other to the norm as a result; (2) there is a remainder to be found in the constitution of the difference between the self and the Other in habit formation, the self and the world in the work of desire and between man and his other woman in 'ethical' life. That the maintenance of a dominant male ethos is based on a gift from women suggests a model of social 'exchange' different to that of contract. And that there is a remainder of difference beyond sexual duality suggests a more dispersed structure of identity than Hegel allows. In this chapter I will explore these two themes further through the idea of the gift as the basis of social exchange. This leads inevitably to a discussion of the relation between deconstruction and an ethics of sexual difference.

JUSTICE AND THE GIFT OF BEING

In chapter 1 I argued, with reference to the surrogacy debate, that contractarian and rights-based ethics rely on a problematic concept of the individual and a dubious model of relations between individuals. These assume, following Locke, that the individual agent is distinct from, and owns property in, their body – property over which the individual ideally rules, property which can be protected by the law in exchange for civil obedience and property which can be exchanged under the security of contract. The individual is said to be naturally autonomous. Free and equal contracts with others, enacted through an exchange of words, are said to be the basis of social relations. I discussed Carole Pateman's objections to this model of social exchange (Pateman

1988). She suggests that, contrary to the claim that contracts are free and equal, they actually constitute a relation of subordination because the party who purchases or 'protects' the property held in another has the power to decide what the other must do to fulfil their side of the contract. Her second objection is to the assumption that the individual agent is male. Women, insofar as our bodies signify womanhood, are parties to a sexual contract only, such that the social contract effectively trades man's civil freedom for woman's social subordination.

I also suggested in chapter 1 that, following Locke, not only does the idea of contract assume a problematic distinction between the individual agent and his or her body but it is also based on a particular model of identity and difference. According to Locke, an entity is self-identical if it has a self-contained origin in time and space. That which is different has a different beginning in time and place (Locke 1975: 328). This principle of individuation allows the claim that the individual already has a self-contained identity (based on the continuity of consciousness) prior to his or her relations with others. A contract (sexual or social) with another can be said to establish a social relationship without affecting the identities of the different parties. So it can be conceded, for example, that the relation between the sexes may involve inequalities under present conditions, but as the identities of each are said to be independent of this relation, it can be assumed that the inequalities can be removed by some legal adjustments to the terms of that relation without questioning the conditions for the production of the ethos of man.

Hegel's model of identity and difference, on the other hand, suggests an entirely different account of exchange in social life. I have already noted Hegel's claims that identity is a product of, and carries within it, its relation to what is different and that the self does not conform to the maxim that everything is identical with itself and that difference is an external relation. Identity and difference do not come before social relations: they are produced within them. My analysis of Hegel's ethics suggests that even when it is acknowledged that identity is constituted in relation to others, man's freedom can still be bought at the cost of women's subordination in the name of the 'common good'. At the same time, the value of Hegel's model of identity and difference lies in the way it exposes what is at stake in the contract model of social exchange: a norm of male body-property is both produced and maintained by containing women's different modes of embodiment outside of the body politic.

In the spirit of a Hegelian model of identity and difference, Marcel Mauss (1967) finds that beneath the artifice of free and equal contracts between self-present individuals lies a form of exchange based on the gift. A gift can be, in theory, anything including an object, a ritual, a woman or a child. Insofar as a gift is of the order of a 'potlatch' (to nourish or consume) its circulation determines the social rank and identity of a society's members. It bestows prestige on the one who receives it and, more important, a moral obligation towards the giver which cannot be repaid in ways other than by maintaining a social bond (Mauss 1967: 6). The power of such gifts to constitute a social bond lies in their spiritual status: transfer of a possession can only establish a social relationship between persons if that possession carries the significance of being part of the personhood of the giver (ibid.: 10). Social-contract theorists may have no argument with this insofar as they assume that part of one's personal property is exchanged through contract (although without any subtraction from one's personhood). However, according to Mauss, if the gift has the power to establish a social relation it is because it *remains* part of the personhood of the giver such that its circulation is one which seeks a return to the place of its birth (ibid.: 19).

There are several ways in which the gift, as a model of social relations, challenges contract theory. First, words (if understood as separable from the speaker) do not constitute a social relation. Rather a social relationship is effected through the gift of part of oneself to another. To the extent that this gift has something to do with the body or the product of its labour, this body cannot be understood in terms of property distinct from the self. It is the self *per se*. Second, the identity of the two individuals is not given in isolation prior to exchange. As what is given is in essence part of the substance of the giver, and as the social identity and status of the recipient is enhanced by the gift, then, contrary to the logic of identity in contract theory, what is constituted through the gift is the social identity of each in relation to the other. Finally, the giver does not pledge obedience with this gift in exchange for its protection. The debtor in this relation is not the giver but the recipient. The gift constitutes the social identity of the parties and an enduring social bond which obligates the recipient to the donor.

This model of social exchange acknowledges that if men accrue social value and prestige it is via a gift of the substantial identity of women. It also finds contracts unethical insofar as it is assumed that the recipient of the gift can pay it out and thereby sign away their obligation to the donor without any further thought of return. (On this model of social relations, surrogacy would not be unethical but a profound expression of the gift. However, the surrogacy contract would be unethical insofar as it allows the social obligation to the birth mother to be paid out. I will return to discuss this further in chapter 6.)

There is, however, a problem with Mauss' model of the gift. He assumes that, under ideal circumstances, the gift can arrive at its destination. In the case of sexual difference women may give to men their social identity and status but the favour is supposedly returned if men maintain their obligation to the women concerned. Marriage, for example, could be said to effect the arrival and return of the gift. It could be acknowledged that, through marriage, a woman gives herself to a man (or, more often, she is given by another man) and he receives with this gift a secure identity and social prestige. On Mauss' model, the gift is effectively returned if the man maintains his obligation to the woman (or her family) through financial and other material support and if, with this, she receives an identity in relation to him. The circulation of the gift, as Mauss sees it, would allow the unity of identity and difference. On this model, no less than on Hegel's, woman's difference is contained within man's identity without acknowledgment of the inequalities this understanding of sexual 'difference' sustains. In other words, in his discussion of the gift, Mauss, along with Hegel, disavows the process whereby man's self-present identity is bought at the cost of the further dispersal of woman's.

Jacques Derrida claims that, insofar as Mauss assumes that the gift is a commodity which can be separated from its donor and returned through a bond of obligation, he is speaking of 'everything but the gift' (Derrida 1992: 180). Mauss, says Derrida, is closer to the economy of contract than he admits. (Or, more exactly, Mauss reverses the terms of the contract, giving the donor more credit, but ultimately retains the contract logic of identity.) According to Derrida, in a contract model of social relations a gift is understood as something which is given by an already constituted self-present donor to an already constituted self-identical donee. But if these are the conditions under which the gift is possible, they are also the conditions under which the gift is destroyed (ibid.: 170). That is, if the donor or the recipient recognises the gift as a thing separate from themselves then a debt will be incurred by the recipient (the gift is no longer a gift as such) and the gift can be annulled by gratitude or some other form of return. The gift is only a gift if it is forgotten or if it goes unrecognised by both the donor and the donee.

Derrida is not just playing with the logic of identity in his analysis of the gift. For him, the gift is the gift of Being (or what I have been calling self-present identity). The impossible structure of the gift is also the structure of Being: Being gives itself in the present on the condition that it is not present (pun intended) (Derrida 1992: 184). That the structure of Being or self-present identity is impossible is based on the (Hegelian) insight that identity is constituted in relation to that which is different, but (contra Hegel) in such a way that the unity of identity is always deferred. This can perhaps be better understood in terms of the operation of what Derrida elsewhere calls *différance*. In the context of his deconstruction of philosophical texts, Derrida defines *différance* as:

> the systematic play of differences, of the traces of differences, of the *spacing* by means of which elements are related to each other. This spacing is the simultaneously active and passive . . . production of the intervals without which the 'full' terms would not signify, would not function.
>
> (Derrida 1981b: 27)

Self-identity cannot be constituted without a production of an interval or a difference between the self and the other. No self-present identity, no relation to Being is generated without this relation to the other. (It is this process of production of identity and difference which constitutes man's habitat as self-contained and privileged and woman's as 'other'.) However, as identity is produced through the other, the 'full' terms so constituted cannot simply refer to or signify themselves. While this production of intervals constitutes an identity as present by separating the present from what it is not (from its other), the

> interval that constitutes it as present must, by the same token, divide the present in and of itself, thereby also dividing, along with the present, everything that is thought on the basis of the present, that is, in our metaphysical language, every being, and singularly substance or the subject.
>
> (Derrida 1982: 13)

As one's identity and social value is produced through a differentiation between the self and the other then the identity of the self is dispersed into the other. *Différance* describes an operation which both constitutes identity and difference and resists and disorganises the totalisation or full presence of meaning, identity or Being. It is the operation of *différance* which insists on the gift: the ultimate dispersal of all identity. And this impossible structure of the gift is such that, if self-present identity is claimed, a debt to the other is incurred.

This idea of the gift has important consequences for an ethics of sexual difference. I have suggested in my reading of Hegel that, when sexual difference is understood in terms of duality or opposition, man's self-presence is won at the cost of woman's. And I have found, beyond that, an incalculable sexual difference – a difference which is suppressed and denied in the accommodation of man's ethos. This sexual difference beyond duality is the realm of the gift. Derrida is reported as saying, for example, that '[a]ll that you can call "gift" – love, *jouissance* – is absolutely forbidden by the dual opposition [between the sexes]' (Derrida 1987: 198). The gift 'produces the identity of the giver and the receiver' as they are given in the relation: identity and difference do not pre-exist the relation, nor does self-present identity flow from it (ibid.: 199). However, insofar as Being, self-present identity, is said to have arrived or been determined in the opposition man/woman then an injustice has been done: the gift of an incalculable sexual difference has been effaced. The possibility of the gift, the possibility of sexual differences beyond opposition and the secondariness of woman, is, for Derrida, the possibility of ethics (following Levinas, with qualification): 'the possibility of ethics could be saved, if one takes ethics to mean that relationship to the other which accounts for no other determination or sexual characteristic in particular' (Derrida 1985: 178). An ethical relation to the other rests on not determining anything about the other's difference ahead of or during one's encounter with them.

Insofar as contractarian ethics assumes that the self is given, present to itself and separate from the other such that the two then negotiate the terms of the secondary relation between them, then the gift (as both the process of production of the interval between them and the other possibilities beyond opposition this implies) is denied. If the self and other are said to be fully constituted and self-contained prior to their meeting and remain unaffected by it, then it can be assumed that the two can meet on equal terms and can, through self-decipherment and interrogation of the other, come to some equitable arrangement. But this meeting is itself a process of production: one which not only suggests a reconstitution of the self in the face of the other but the irreducibility of the self to the self and the other to me. In assuming autonomous, self-present identity and the potential reduction of the other to the self, contractarian ethics perpetuates a violence against the other where the other's undecidable difference is effaced.

Further, in an environment where autonomous self-present identity is granted to men such that women are already constituted as other to privileged identity, it is unlikely that, upon their meeting, the woman's interests can be acknowledged, represented or articulated in the 'dialogue' between them or by the rules that govern their relations. For example, as discussed in chapter 1, the interests of a pregnant woman cannot be represented in such a paradigm. Either the contract with the father of conception (governing her pregnancy) is deemed to be sexually neutral, in which case her experience is subsumed under a male norm to her disadvantage. Or it is granted that her sex is different and her participation in social exchange is disallowed on that basis. Here the woman's difference is taken to be locatable, understood and deemed other to that which is required for just and equitable social relations. Declaring either that the sexes are the same or that woman is different and therefore other to the communal ethos denies the operation of the gift in the constitution of identity and difference.

These difficulties are not overcome by simply paying attention to context, by attempting to subsume the other's particular context within general rules or by asking the other's opinion. Insofar as the other is thought to be transparent or is accommodated within norms already in place, the other's irreducible difference, the gift of new possibilities for existence, is denied. Any 'dialogical' or 'contextual' ethics which is based on such assumptions may acknowledge a constitutive relation between generalities and context and between self and other, but risks repeating the effacement of differences apparent in Hegelian dialectics:

> when sexual difference is determined by *opposition* in the dialectical sense . . . one appears to set off a war between the sexes; but one precipitates the end with victory going to the masculine sex. The determination of sexual difference in opposition is destined, in truth, for truth; it is so in order to erase sexual difference. The dialectical opposition neutralises or supersedes . . . the difference.
>
> (Derrida 1985: 175)

Or, as Drucilla Cornell suggests, more generally: '[t]he shutting in of context, the denial of new possibilities yet to be imagined, is exposed [by the operation of *différance*] as political, not as inevitable and, more importantly, as unethical and ultimately unjust' (Cornell 1991: 109).

Cornell, following Lyotard, defines injustice as 'damage accompanied by the loss of the means to prove the damage' (Cornell 1991: 110). An ethics which normalises, an ethics which assumes self-presence and posits difference as a secondary relation and which, following this, attempts to neutralise differences (in the name of equality), or an ethics which claims to have included the other's specific context, damages the other by reducing the other to the same. As Cornell suggests, in a culture 'scarred by gender hierarchy', a theory of justice which assumes or seeks sexual neutrality ignores the harm done to women in the production of this scar and doubles the injustice by insisting that harm be translated into the terms of a system which does not recognise it (ibid.: 110, 114). For Cornell, justice requires the deconstruction of this system of injustice, an interrogation which opens up the operation of the gift, *différance* or that deferral towards other possibilities which resists normalistion. Or, as Derrida puts it, justice is not a matter of neutralising differences but, rather, justice requires us 'constantly to maintain an interrogation of the origin, ground and limits of our conceptual, theoretical or normative apparatus surrounding justice'.[1]

Derrida's deconstruction is an 'ethics' based on the gift or recognition of the irreducibility of the 'other' to 'me' and is, therefore, an 'ethics' which, like Foucault's, exposes the unethical basis of normalising codes of justice (Harvey 1986: 227). As such an ethics works against universalist ethics which assumes self-present identity and a male ethos as a norm, it should be consistent with an ethics of sexual difference. However, while the gift promises the possibility of incalculable, and hopefully more equitable, sexual differences it does not necessarily rest easy with feminism. The unease arises from the haste by which the move is made from an interrogation of the conditions of the constitution of (man's) self-present identity to a celebration of incalculable sexual differences. The more hasty the move from interrogation to celebration the more likely something will be forgotten, namely the negative effects on the bodies of women of the privileging of man's self-present identity. I turn now to consider this difficulty through an examination of the conditions under which this idea of the gift may work towards justice for women and the conditions under which it does not.

DECONSTRUCTION AND SEXUAL DIFFERENCE

The gift of incalculable sexual differences does not emerge from injustice unaided but, as Cornell and Derrida suggest, through the labour of interrogation or deconstruction of normalising social discourses. And deconstruction, by opening discourse to the gift, suggests that all identity is ultimately dispersed. However, as analyses in previous chapters indicate, in discourses which privilege a self-present male ethos, it is women who are associated with dispersal and undecidability. Not only is woman's identity taken to be dispersed, but this is viewed as a pathological condition. It is therefore not surprising that Derrida increasingly associates the impossibility of identity (which for him is not pathological) with the figure of woman. The positive effect of this association is that the metaphor of woman is given a special status as that which disrupts dominant, normative identity (Grosz 1989: 34). The association also positions sexual difference as the central issue in an ethics of difference. But the association between 'woman' and the operation of a difference which exceeds binary oppositions is not without its problems. There are at least two problems raised by interpretations of Derrida's approach to difference. Both, it seems to me, arise from his apparent neglect of the question of embodiment and a subsequent difficulty, on the part of his readers, in locating embodied women in the operation of *différance*.

The first problem arising from the association between 'woman' and *différance* relates to how we are to understand the relationship between the gift (*différance* or dispersed identity) on the one hand and normalising social discourses (including ethics) on the other. Some postmodernists would seem to assume, in announcing the arrival of the gift, an absolute disjunction between the two. That is, it is often assumed that, because the work of Derrida and other 'poststructuralists' suggests that one's identity is always dispersed in another's, then it is enough to acknowledge this 'fact'. This is the problem of moving too quickly to the gift by declaring an end to hierarchical oppositions between man and woman and a rejection of the normalising discourses which maintain this inequity.

Craig Owens, for example, in his influential paper 'The Discourse of Others: Feminists and Postmodernism', claims that the work of Derrida, Ricoeur, Lacan and Lyotard points to what is symptomatic of contemporary culture: decentredness and a loss of mastery (Owens 1985: 67). He celebrates postmodern art which 'testifies to a *refusal* of mastery', in particular work by women which questions identity and authorship while avoiding '"positive" images of a revised femininity' (ibid.: 68, 71). But to simply declare that man's identity is dispersed does not make it so. Such declarations of dispersed identity rely on assuming an absolute distinction between the gift, on the one hand, and a normalising ontology of self-presence on the other, as if the gift signifies a void of meaninglessness which is prior to, covered over and ultimately unaffected by normalising social discourses. The distinction between dispersed identity and self-presence allows the assumption that you can just reject self-presence and dispersed identity will emerge. Yet, as I have indicated, in outlining the operation of *différance,* the two cannot be separated in this way: the gift opens within the very process of production of self-presence which attempts to deny it. There is no loss of mastery here; to claim otherwise is to miss the point of the philosophical tradition which challenges an ontology of presence, and this is to miss the point of deconstruction. Deconstruction is testimony to mastery and to the material effects of the conceptual oppositions which hold the master in his dominant position (if there were no injustice and no claims to self-presence, deconstruction could not proceed). And, despite Owens' best intentions, his celebration of a refusal of female identity, in a climate where masculine 'self-mastery' is still the rule, teeters dangerously close to anti-feminism.

Jean Graybeal (1990) also appeals to the distinction between dispersed identity and normalising discourses which assume self-presence but unlike Owens her approach is decidedly feminist. One place where she finds hope for other possibilities for women, beyond subordination to men, is in the primordial dispersed structure of human being-in-the-world (*Dasein*) as Heidegger outlines it in *Being and Time.* In its least problematic formulation this dispersed structure refers to the impossible structure of identity which Derrida calls the gift (and which he often evokes in Heidegger's name) and which Heidegger refers to as Care or ontological difference (the difference between being and Being or the openess of our existence to undetermined possibilities). More problematic is the way Heidegger tends to oppose this dispersal of identity to 'inauthentic' living in the 'they' (*das Man*). Inauthentic living in the 'they' refers to our tendency to measure and understand ourselves in terms of those others we encounter (Heidegger 1962: 163–6) and in terms of 'idle talk' or interpretations of beings which assume presence (ibid.: 212, for example). Included in the discourse of the 'they' is the measurement of sexual difference in terms of duality (Heidegger 1984: 136–9).[2] This process of comparison with others is normalising because through it we effectively allow others to dictate our modes of existence, we take these modes of being for granted and allow our own possibilities for existence to dissolve into public anonymity. Against this inauthentic existence Heidegger posits authenticity: a call back to a more primordial dispersed structure where other possibilities are kept open. At its most extreme, the distinction between inauthenticity and authenticity implies that the dispersed structure of our existence is prior to and covered over by misinterpretation and the normalising discourses of the 'they', and that we can recover our other possibilities through withdrawal from the 'they' (Heidegger 1962: 294, for example) and through silent rejection of its interpretations (ibid.: 164, 342).

Insofar as Heidegger equates the keeping open of ontological difference with authentic solitude his formulation of the dispersed structure of identity is problematic.[3] But it is to this notion of authenticity and its opposition to inauthenticity that Graybeal appeals. With some labour, Graybeal finds the 'feminine' in Heidegger's notion of authenticity. She equates the language of the 'they' with Julia Kristeva's notion of the 'symbolic' dimension of language: that which universalises everything under the Law of the Father and hence that which privileges men (Graybeal 1990: ch. 5). The call to authenticity, she argues, is a collapse of signification, and therefore of the identity of the 'they-Self' (ibid.: ch. 6). This return to 'authenticity', this dissolution of inauthentic identity, is akin to the effect of the 'semiotic' dimension of language which opens existence to other possibilities beyond those dictated by the 'symbolic'. That the semiotic interruption of the symbolic (the return to authenticity) is 'feminine' is evidenced, according to Graybeal, by Heidegger's use of the fable about the Goddess Care (Care is another name he gives to the dispersed structure of *Dasein's* existence) and by what seems to be an assumption that dispersal of identity is, by definition, either feminine or maternal.

We may agree that women's identity is dispersed as an *effect* of a language which insists on man's self-presence, but to separate this dispersal of identity from the language of self-presence is problematic.[4] By relying on the equation between authenticity and the 'feminine' and privileging authenticity as a way of opposing the normalising measurement of sexual difference by the 'they', Graybeal implicitly privileges silence or withdrawal as feminist practices of resistance. Or, in relation to the status Heidegger gives to the poetic in his later work, Graybeal's suggestion is that we accept and delight in being always divided and dispersed in ourselves rather than projecting this 'dividedness' outward (on to women) (Graybeal 1990: 160).

Heidegger's notion of authenticity (embraced uncritically) is not enough to evoke other possibilities for women, nor is the related call to keep our dividedness to ourselves. If self-identity is divided or dispersed this is the result of being-with-others, of being constituted in relation to others under the guidance of an ontology of presence which deems woman to be man's other. Appeals to a feminine dispersed identity, as if it were outside of existing relations between the sexes, are not only hollow but risk entrenching the association between women and that which is said to be pathological. If there are sexual differences beyond sexual duality then this can only be opened by an interrogation of normalising interpretations of sexual difference in terms of presence rather than by withdrawing from the other or the 'they' into authenticity. I am suggesting that an ethics of sexual difference needs to 'deconstruct' the ontology of presence and the discourse of the 'they' rather than rely on what may emerge from its refusal (assuming refusal were possible).

As Gayatri Spivak suggests, the value of deconstruction lies not in its announcement that the self is decentred, but in its suggestion that *the subject is always centred* and at home and that the means of this centring delimits others as not at home (Spivak 1989: 214).[5] Simply announcing the death of identity leaves in place, to do their work more effectively, those discourses and practices which constitute and privilege the habitats of some men. To the extent that postmodern thinkers find a solution to injustice in solitude, their work does nothing to change the terms of injustice. The coherence of man's claims to autonomous self-presence, which regulate our social and political structures, can be called into question, not by declaring them non-sense, but by scrutinising the exclusions upon which such claims are based. Deconstruction undertakes this questioning.

Appeals to authentic solitude as a way of opposing normalisation are not restricted to a dependence on an originary dispersed identity in the sense just discussed. Foucault's aesthetics of existence, outlined in chapter 2, also seems to require a kind of solitude or at least a distancing of oneself from the normalising social moral code and from the other. That new modes of existence may emerge from this distancing appears to depend, in Foucault's case, on a pre-social body of disorganised pleasures. From what I have been arguing here, Foucault's aesthetics of self may be found wanting insofar as it ignores, and possibly perpetuates, the power play operating in conceptual oppositions which generate and privilege man's identity. I want to briefly examine Foucault's appeal to solitude, not just to mark its difficulties and its difference from deconstruction, but to mark it off from his earlier work on the normalisation of bodies. It is Foucault's attention to embodiment and the work of disciplinary power, rather than his aesthetics of self, which is useful for an ethics of sexual difference.

Foucault points to a difference between his work and that of deconstruction in an interview where he raises the question of the status of the *hypomnemata* (which he defines copybook or notebook) in Plato's thought. This is the question of the relation between writing (or self-representation) and the constitution of self. He begins by claiming that 'Current interpreters see in the critique of *hypomnemata* in the *Phaedrus* a critique of writing as a material support for memory' (Foucault 1984: 363). While he does not mention Derrida by name, this seems to be referring to Derrida's analysis, in *Dissemination*, of Plato's critique of writing as a replacement for memory (Derrida 1981a: 95–117). Plato, in a manner not unlike Locke, locates self-presence in thought (soul) and the continuity of self in the internal workings of a reliable memory. Those who represent their thoughts through writing will, according to Plato:

> cease to exercise their memory and become forgetful; they will rely on writing to bring things to their remembrance by external signs instead of their own internal resources. What you have discovered [in writing] is a receipt for recollection; not for memory.
>
> (Plato 1973: 275)

Derrida is critical of Plato's assumption that behind this supposedly unreliable mode of representing the self (a writing which separates the self from its representation) is a self-presence immediately signified by memory.

Foucault proposes what he claims to be an alternative interpretation of the relation between writing and the self in Ancient Greek thought. *Hypomnemata,* he suggests, were not viewed as re-presentations or accounts of the self; rather, they formed a technique for collecting and reassembling the 'already said', the 'fragmentary *logos* transmitted by teaching, listening, or reading' for the purpose of the *constitution of oneself* (Foucault 1984: 365).

What Foucault is doing here is reversing the usual order between the 'inside' of the self and its representation, claiming that representation, as writing, does not signify to the 'outside' that which is already 'inside' the self. Rather, this writing constitutes the self. However, in one sense, this does not mark a difference from Derrida's analysis. Derrida attempts to show the impossibility of Plato's dream of a memory without a sign, presence without absence in some form of writing, a self without re-presentation. In short, he demonstrates that writing, as a sign or re-presentation of the self is not supplementary; writing constitutes the self.

But there is something more to Derrida's analysis which may signal his difference from Foucault: the impossibility of an aesthetics of self which escapes its debt to the other. According to Derrida, the *hypomnemata* is constitutive of self because:

> What is is not what it is, identical and identical to itself, unique, unless it *adds to itself* the possibility of being *repeated* as such. And its identity is hollowed out by that addition, withdraws itself in the supplement that presents it.
>
> (Derrida 1981a: 168)

Self-representation is a necessary condition of the constitution of self. But the necessary condition of self-representation is the absence of what is presented. Hence the structure of the constitution of self, whatever the technique, is that the self will be divided from itself, finding within its identity a trace of its other. If this dispersal of identity is denied, if it is assumed that the self can 'gather together' into itself that which grounds its identity, then, as I have suggested in my account of the gift, a debt to the other is incurred and denied. However, if Foucault is guilty of reclaiming self-presence in his aesthetics of self, this is only as a result of forgetting his own, earlier account of how the embodied self is constituted by disciplinary power within hierarchical relations with others. Here he does acknowledge that the value and identity of the self as a corporeal artefact is defined 'in relation to all other differences' (Foucault 1979: 183). This work on body/power is worth retaining, against an apparent neglect of embodiment within deconstruction. It is this problem with deconstruction I will now explore.

Besides the tendency of some to hastily announce the arrival of the gift, the second problem for deconstruction as an ethical practice comes from critics of Derrida who suggest that his notion of *différance* is either a co-option of the feminine (Jardine 1985: 207, for example) or an oppressive feminisation of a difference which has nothing to do with 'real-life women' (Braidotti 1989: 89; 1991: 103). While critical, these informed notes of caution deserve more serious attention, I feel, than the celebration of dispersed identity which infects Owens' kind of postmodernism. Ironically, however, both the criticism that *différance* ignores the reality of women and the claim of a loss of self-mastery come from the same assumption: that *différance* implies an idealist or non-materialist operation of representation.

To say that *différance* has nothing to do with real-life women is to say that the difference which exceeds the opposition between man and his other has nothing to do with the way women live their embodiment. The validity of this criticism (and the claim of co-option) depends on the status of the bodies of women (and those of men) in deconstructive discourse. That is, it depends on what we understand the relation to be between the material, lived experiences of women and both the metaphor of the 'other' woman (the idea of woman against which the subject affirms his self-presence) and the metaphor of woman as a model of undecidability. I suggest that both 'metaphors' are constitutive of the bodies of women. That is, the bodies of women (real-life women), our habits, desires and experiences are constituted by social discourses and practices which position women as other to privileged identity. But insofar as there is an operation of a sexual difference which exceeds this duality, then 'real-life women' are suspended between that which upholds a dominant male ethos and that which remains other. This excess too is lived by women as part of our material reality. Or, as Vicki Kirby puts it: '*différance* is better understood as the condition of possibility, the spacing, in which women's embodiment is inscribed' (Kirby 1991: 100).

Différance, while not a substance, does nevertheless indicate the uncertainty apparent in the production of substances. For example, Derrida follows a comment about the possibility of a less discriminating sexual difference beyond binary difference, by this remark:

> I would like to believe in the multiplicity of sexually marked voices . . . [a] mobile of non-identified sexual marks whose choreography can carry, divide, multiply the body of each 'individual', whether he is classified as 'man' or as 'woman' according to the criteria of usage.
>
> (Derrida 1985: 184)

The 'play' of *différance* which constitutes, and defers the closure of, sexual duality, is not outside the bodies of men and women. Just as women materially bear the burden of concepts which write us out of the social text, our habitats, so constituted, are no less self-contained than those of men. Women's bodies are also marked by an opening towards other possibilities. This opening is as real as the embodied concepts which seem to close it.

Despite these indications, the status of embodiment, and of materiality in general, remains precarious at best in interpretations of Derrida's work. At worst, the body of woman has been entirely written out of deconstructive discourse. Both critics and followers of Derrida tend to interpret his proposition 'there is nothing outside of the text' (Derrida 1974: 158) to mean there is no referent, no reality, only fictions, metaphors, multiple meanings. It is therefore not surprising that '*différance*' is understood, paradoxically, as a metaphor of 'woman' which has nothing to do with real-life women. Yet Derrida, at least, does not forget the material referent:

> It follows that if, and in the extent to which, *matter* in this general economy designates, as you said, radical alterity (I will specify: in

> relation to philosophical oppositions), then what I write can be considered 'materialist'.
>
> (Derrida 1981b: 64)

As Cornell argues, Derrida's work is about the productive power of language (Cornell 1991: 28, 104). And, as Robyn Ferrell suggests, *Il n'y a pas de hors-texte* would be better translated as 'there is no outside to the text' (Ferrell 1993: 126). This would better indicate Derrida's conviction that 'there is no "elsewhere" that escapes the rigours of representation' and would better signal the deconstructive project as one which contests, not reality, but the philosopher's desire for 'the pure outside he seeks, uncontaminated by his own subjectivity' (Ferrell 1993: 126, 130).

That Derrida thinks representation is a material production is indicated in his reading of Hegel's theory of the sign which I discussed in chapter 3: here the body of the sign is transformed in signification to represent something it is not. *Différance* does not then indicate a pool of differences which have yet to find their material signifiers: it does not denote a void of meaninglessness in opposition to the full presence of meaning. *Différance* is however testimony to Derrida's opposition to idealism, to the (Hegelian) assumption that matter, or any body, is reducible to a concept it comes to represent. There is always a material excess in the production of identity and difference. But, at the same time:

> If I have not very often used the word 'matter', it is not, as you know, because of some idealist or spiritual kind of reservation. It is that in the logic of the phase of overturning this concept has been too often reinvested with 'logocentric' values, values associated with those of thing, reality, presence in general, sensible presence, for example . . . In short, the signifier 'matter' appears to me problematical only at the moment when its reinscription cannot avoid making of it a new . . .'transcendental signified'.
>
> (Derrida 1981b: 64–5)

Insofar as *différance* evokes a material remainder to the economy of representation which confines woman to the position of man's deficient other, then it indicates that the bodies of women are open to other possibilities. But this openness of embodied experience is not an objective reality or a transcendental signified prior to, and outside of, the work of signification. This perhaps marks another difference between a feminist deconstructive project and Foucault's aesthetics of self (insofar as Foucault relies on a body outside of signification).

While offering this defence of the relevance of deconstruction to an ethics of sexual difference, it would be foolish to assume that the tendency for both critics and disciples of Derrida to write embodiment out of deconstruction is based simply on a misunderstanding. The preconditions for that misunderstanding are no doubt there. Correction then is not a matter of simply pointing to relevant passages in Derrida's work or that of deconstructive feminists, as I have done here. Nor do I think that the difficulty is overcome by 'giving body' to the 'irreducibility of the feminine other' by affirming it through metaphors (Cornell 1991: 100–1). This irreducibility of the other is already embodied insofar as it is an open embodied effect of metaphor, of naming woman as other. If the embodied experiences of women are to be written back into deconstruction, this must be done not by insisting that the gift has arrived as embodied or not, but by interrogating the material conditions of its erasure. So, if poststructuralism is to deliver an ethics of sexual difference, the problematic of the constitution of place must account for, if not emphasise, the conditions for production of sexed bodies. One's body cannot be separated from the discourses which inform it with meaning and value and which it (almost) comes to represent. At the same time, it is also necessary to recognise that, insofar as one's body is this sign which is the self, it is divided in itself. And, as the body always refers beyond itself, its production is incomplete.

There are three main indications for an ethics of sexual difference which arise from this conclusion. First, against a tendency to forget materiality in some modes of postmodernism, the production of sexed bodies is inextricably caught within the production of the other and within the discourses which describe as they constitute sexual difference. Second, insofar as the production of difference is incomplete, sexed bodies are always open to other possibilities beyond those which position woman as other to man. Even though there is no 'outside' the social text, this does not preclude the possibility of change. Finally, against the tendencies apparent in the work of Foucault and postmodernists such as Owens, an aesthetics of self which claims not to refer to a social code or not to 'other' the other woman, leaves both the social code and the other in place.

As alterity is inherent in the structure of any sign of the self, an aesthetics of self which forgets its debt to the other forgets its complicity in the subjection of the bodies of women. Or, as Gayatri Spivak puts it:

> The solution is not merely to say 'I shall not objectify'. It is rather to recognise at once that there is no other language than that of 'objectification' and that any distinction between 'subjectification' and 'objectification' is as provisional as the use of any set of hierarchical oppositions.
>
> (in Derrida 1974: lix)

This is not, of course, an apology for 'objectification', oppression or exclusion of the other. In challenging the apparent indifference of normative discourses it is important to recognise that the representation of sexual 'difference', which upholds a male norm, depends on the 'objectification' of women. But to assume one can simply sidestep this process, rather than recognising and interrogating one's complicity with it, is to return to the assumption of self-presence and its concomitant effacement of alterity. If it is possible to imagine a place for sexual differences, beyond the economy of exchange which assumes self-presence and sameness, it is not a place outside the social text which constitutes and houses the bodies of women and men.

I have been suggesting that in evoking the possibility of the gift, of sexual differences beyond the determinations of difference which keep women in a subordinated position, we must also deal with the material effects these determinations have on women. This cannot be achieved by ignoring or withdrawing from our everyday determinations of difference and their metaphysics of presence. In the following chapters I return more directly to address the question of embodiment but with the added insight that sexed bodies are constituted within an economy of identity and difference which limits possibilities for women. I begin with Nietzsche because his philosophy pays attention to both these considerations: the social constitution of the embodied self and the idea that this occurs within a field of relations with others and within an economy of hierarchical conceptual oppositions. Nietzsche's 'ethics' is not without its problems when it comes to sexual difference. However, the reading I will provide suggests that insofar as he forgets either of the considerations mentioned, he perpetuates an injustice against women. To the extent that this occurs, it is possible to expose what is at stake in an ethics of sexual difference.

Chapter 5

Nietzsche on sexed embodiment

The two themes I have claimed are necessary to an ethics of sexual difference, understood as the problematic of the constitution of one's ethos, are: the itinerary of the social constitution of sexed bodies and the conditions for the production of sexed identity and difference. Both themes come together in Nietzsche's ontology and his critiques of Christian, utilitarian and contractarian ethics. In the following reading of Nietzsche's philosophy I will draw out these themes, not just as a critique of how ethics is usually undertaken, but also as a warning against that tendency discussed in the previous chapter: the tendency to move too quickly to the gift of dispersed identity by either announcing its arrival or by positing a solitary aesthetics of self.

Nietzsche has his own aesthetics of self: a project for embodied self-creation which works against the normalising effects of moral codes. It is Nietzsche's insight that perspectives are embodied which works against the assumption, apparent in some contemporary approaches to the gift, that the opening towards other possibilities for existence is divorced from material reality. And it is his insight that the body as one lives it is both a cultural artefact and the site of change which Foucault develops in his account of disciplinary power and an aesthetics of existence. However, there is a significant difference between Foucault's and Nietzsche's ethics of difference: Nietzsche acknowledges that the division within the self necessary for creative self-formation is predicated on a relation to the other. This relation is productive: the reconstitution of the self's identity implicates others. Insofar as Nietzsche recognises that an aesthetics of self is built upon the other, his ethics has important consequences for an ethics of sexual difference. But, as I will go on to argue, insofar as he forgets the other, his aesthetics of self reproduces the exclusion of others and remains complicit in the subjection of the bodies of women.

THE BODY AND SELF-FORMATION

Central to Nietzsche's concept of self, and a point often overlooked by 'hyperreal' postmodernism, is his recognition that the problematic of the constitution of place is a question of the social constitution of embodiment. In *Thus Spoke Zarathustra* he claims that 'body I am entirely, and nothing else; and soul is only a word for something about the body' (Nietzsche 1978: 34). In contrast to assumptions that the self's identity can be reduced to consciousness and that the mind directs the body, Nietzsche claims that the body is what compares and creates and that thought and the ego are its instruments.

He is not suggesting that the body is an a-social fact in charge of operations. And, while 'in man *creature* and *creator*', matter and sculptor, are united (Nietzsche 1972: 136), it is not consciousness (transcendental or individual) which makes a man out of matter. Rather, the body like any 'thing' is the sum of its effects insofar as those effects are united by a concept (Nietzsche 1967: 296). The 'body is only a social structure composed of many souls' where 'soul' refers to a corporeal multiplicity or a 'social structure of the drives and emotions' (Nietzsche 1972: 31, 25). So, for Nietzsche, one's place in the world, one's desires, habits and potentialities are determined by the concepts which govern the structure of the social world and which sculpt the body accordingly – a body which is a 'unity as an organisation' and is therefore a 'work of art' (Nietzsche 1967: 419).

How the self is made as a social structure is first a question of how the body is unified through social concepts. Second, and related to this process of unification, is the question of how thought and the ego are instruments of the body. Taking the second claim first, the body is the locus of pleasure and pain (which are always already interpretations) and thought is a reflection on pleasure and pain. To quote Nietzsche:

> The self says to the ego, 'Feel pain here!' Then the ego suffers and thinks how it might suffer no more – and that is why it is *made* to think.
>
> The self says to the ego, 'Feel pleasure here!' Then the ego is pleased and thinks how it might often be pleased again – and that is why it is made to think.
>
> (Nietzsche 1978: 35)

Thought is about the projection of bodily experience (pleasure and pain) into the future: the conscious subject is an effect of temporalising the body.

The target for much of Nietzsche's critical attention is the manner in which experience is unified and the body temporalised by the moralities of modernity. Here, the embodied self is constituted by social concepts which discourage difference, creativity and change. His account in the second essay of *On the Genealogy of Morals* begins with the idea that the unification of any body relies on the operation of memory and forgetting. 'Forgetting' is the incorporation of bodily effects before they become conscious and a making way for new sensations by allowing one to 'have done' with the old (Nietzsche 1969: 58). While this not-remembering is necessary for the constitution of any self as present, the making of the modern moral subject, the individual who is responsible for his or her acts, requires a faculty which opposes forgetting – memory.

While Locke understands memory as a given mode of recollection and self-unification, Nietzsche describes how such a memory is constituted through social and moral discourses of modernity. This is a particular kind of memory: a memory which unifies a selection of activities, events, experiences and effects such that they belong to one person (Nietzsche 1969: 58). This memory makes the self constant and apparently unchanging through time by projecting the same body into the future. The operation of memory and forgetting unifies experience in another sense – it makes different experiences the same. What is remembered is not just an experience but a socially prescribed mode of interpreting that experience. As Nietzsche explains in *Twilight of the Idols* (Nietzsche 1968: 50–3), effects and events are incorporated by interpretation using prevailing moral norms and the concept of cause. Unpleasant feelings are said to be caused by actions considered undesirable. Pleasant feelings are said to arise from good or successful actions. Hence, 'everything of which we become conscious is arranged, simplified, schematized, interpreted through and through – . . . pleasure and displeasure are subsequent and derivative phenomena' (Nietzsche 1967: 263–4).[1] So even forgetting defined as having done with an event involves two processes. First, effects are divided into those which are written into the body and those which are not. Second, events which are incorporated and upon which we reflect are divided into a cause and an effect where the effect is pleasure or displeasure and the cause is interpreted according to social moral norms. Then, when encountering a new event or effect, the memory 'calls up earlier states of a similar kind and the causal interpretations which have grown out of them' (Nietzsche 1968: 51). New experiences are subsumed under habitual interpretations making every experience a fabrication (Nietzsche 1972: 97).

The individual is not the author of this dutiful memory – it is created through what Nietzsche calls the 'mnemotechnics of pain' (Nietzsche 1969: 61): techniques of punishment which carry social norms and moral values. 'Body I am entirely' insofar as my conscience, sense of responsibility and uniformity are created by an ordering of sensations, and a projection of the body into the future through a social disciplinary system. This ensures not only that an individual's experiences are consistent over time, but that as we are subjected to the same moral values we will all have 'our experience in common' (Nietzsche 1972: 186). Forgetting in conjunction with a selective memory becomes a social instrument of repression against the dangers of inconsistency and variation. A society which favours consistency and conformity discourages us to leave our embodied place behind.

Contrary to the individual assumed in contractarian and utilitarian ethics, Nietzsche proposes that the individual is a cultural artefact whose existence is a product of the exclusion of other possibilities for one's embodied place in the world. But this account leaves Nietzsche with a problem shared by Foucault and those who find self-mastery and universal values oppressive: how can change be effected given that the self is the result of a socially informed material process of production? How can different possibilities for one's embodiment be opened, without assuming the possibility of stepping outside either one's present body or one's social context? It is Nietzsche's concept of a distance or division within the self which addresses this apparent impasse.

The body which conforms to a uniform mode of subjection is one which acts out a social role imposed upon it.[2] In contrast to this actor, Nietzsche privileges a process of self-fabrication with the artistic ability to stage, watch and overcome the self according self-given plan (Nietzsche 1974: 132–3). He draws on two features of art and the artist to characterise creative self-fabrication (ibid.: 163–4). The first is the suggestion that the self, like any artefact, is an interpretation, perspective or mask. Second, the relation between artists and their art illustrates the point that creating beyond the present self requires that we view ourselves from a distance in an image outside ourselves. Leaving behind the influence of social concepts which restrict our possibilities in the world requires treating one's corporality as a work of art.

The distinction Nietzsche makes between the self as artist and the image or spectacle staged beyond the present body could imply a unique, extra-social invention. But at a less ambitious level it suggests that you are never identical with yourself. Nietzsche sometimes refers to this division or difference within the self as the 'pathos of distance':

> that longing for an ever increasing widening of distance within the soul itself, the formation of ever higher, rarer, more remote, tenser, more comprehensive states, in short precisely the elevation of the type 'man', the continual 'self-overcoming of man', to take a moral formula in a supra-moral sense.
>
> (Nietzsche 1972: 173)

What Nietzsche is suggesting here is that the ability to move beyond oneself hinges on a relation within the soul (where the soul is something about the body). A distance or difference within the self, between the present self and an image of self towards which I aspire, is necessary for change to be incorporated in the constitution and enhancement of the bodily self. We should not confuse the artist and his work, says Nietzsche, 'as if [the artist] were what he is able to represent, conceive and express. The fact is that *if* he were it, he would not represent, conceive, and express it' (Nietzsche 1969: 101). The self as a work of art is never the same as the self that creates it, not because the self as artist is the true or essential self in contrast to a false, unique, extra-social image projected. Rather, the image which the artistic self creates is a moment beyond the present self which creates it. The difference, or distance, between the two is a precondition to representation which for Nietzsche is always self-representation.

In *Thus Spoke Zarathustra*, Nietzsche accounts for this distance within the self in terms of a process of self-temporalisation of the body which subverts the notion of linear time assumed in contract theory and other normalising social discourses. Unlike the 'last man', who views himself as the essential and unchangeable endpoint of his history (Nietzsche 1978: 202), the 'overman' views himself as a moment. He risks his present self or, as Nietzsche puts it, 'goes under' (ibid.: 14–15). But, unlike the 'higher man', who, in a manner similar to some versions of the dispersed self, affirms the future by negating the past and skipping over existence, thereby changing nothing (ibid.: 286–95), the overman risks himself by 'willing backwards': 'To redeem those who lived in the past and to recreate all "it was" into "thus I willed it" – that alone I should call redemption' (ibid.: 139). Self-creativity is not a matter of declaring oneself born again by simply reaching for a new part to play: it requires working on oneself. The overman then is the self that is a moment which temporalises itself by recreating its past as a way of projecting itself into the future. This self-temporalisation produces a distance or difference within the self. This idea, that the bodily self is reproduced differently as it is temporalised through the production of a distance within the self, is contrary to the assumption central to contractarian ethics, that the self remains the same throughout changes to one's embodied existence.[3] The structure of the moment which is the present self (ibid.: 157–60) is one where the self exceeds its present self rather than one where the self is self-present and self-identical. Man is 'an imperfect tense' (Nietzsche 1983: 61): his past is never complete in relation to his present.

The distancing effected by making the moment one's own is not a state of mind: it 'creates a higher body' (Nietzsche 1978: 70) – the bodily self is reproduced differently. Or, as Judith Butler puts it with special reference to the reproduction of sexed identity: embodied identity is reconstituted in time through the repetition of acts, but as there is always difference in repetition this is a 'stylized repetition of acts' (Butler 1990: 140). That self-representation involves difference is also apparent in Nietzsche's use of the metaphor of pregnancy to characterise the artistic self.[4] The overman 'begets and bears' (Nietzsche 1972: 113) a future self which is beyond and different from himself. This process of self-formation is not a simple rejection of one's embodied ethos. Nietzsche's formulation of a distance within the self re-opens what is denied by social discourses which, in assuming an unchanging subject over time, assume that 'what is does not *become*' (Nietzsche 1968: 35). This assumption of identity as sameness is an 'escape from sense-deception, from becoming, from history' (ibid.). The history which conformity disavows is the process of incorporating new experiences and shedding the old, reconciling conflicting impulses, the ongoing process of corporeal self-fabrication according to concepts one has inherited and cultivated (Nietzsche 1974: 96–104; 1972: 269–71).

SELF-FORMATION AND THE OTHER: THE CREDITOR/DEBTOR RELATION

Nietzsche's model of creative self-fabrication, like Butler's and Foucault's which follow it, allows for resistance to the normalising ethics of modernity. But, like Foucault's, it remains an uneasy formulation with regard to the status of the other. Nietzsche often speaks as if the distance within the self effected by making the moment one's own is generated by the self alone: creative self-fabrication is often presented as an autonomous, self-contained project. Yet, in 'Schopenhauer as Educator', for example, Nietzsche suggests that rather than finding ourselves within ourselves we are more likely to find ourselves outside ourselves, that is, in our effects, in 'everything [which] bears witness to what we are, our friendships and our enmities, our glance and the clasp of our hand, our memory and that which we do not remember, our books and our handwriting', in the objects we love (Nietzsche 1983: 129). In other words, the self is not just divided between the remembered and the forgotten, the future and the past, but between the self and the other. There is something about our relation to others which determines both our embodied place within social relations and the division within the self necessary for change. Hence, contrary to some postmodern formulations of a dispersed self who does not 'other' others, creative self-fabrication must implicate others in some sense.

Nietzsche's genealogies of justice and punishment typically reveal the ways in which others are involved in the constitution of one's ethos. In these we find a tension between understanding the self/other relation in terms of a contract between self-present individuals and understanding it in terms of the uncertain operation of the gift. The most fundamental social relation is, he claims, the creditor/debtor relation where 'one person first *measured himself* against another' (Nietzsche 1969: 70). This relation of measurement is a precondition to punishment: the inflicting of pain under the eye of the law. While techniques of pain have become a way to create that memory necessary for conformity, Nietzsche suggests that inflicting pain on another was 'originally' a way of recovering a debt. And this involved evaluating different parts of the body to ensure that the pain inflicted was equivalent to the debt owed (ibid.: 62–5). Under such a system, evaluation is of the body and operates by mutual agreement. Debts can be repaid through the body via an arrangement between creditor and debtor. If this arrangement is to be understood in terms of contract, then it is a contract written in blood, not words. It is a relation where the status of one party is built upon the body of another.

But there is much to suggest that the creditor/debtor relation is not at all a contract between self-present individuals. While Nietzsche speaks here as if there is an original difference between debtor and creditor, the debtor and creditor are constituted as such through the process of corporeal measurement, not prior to it. As determining values, establishing and exchanging equivalences is the most fundamental social relation, then evaluation of one's own body in relation to another is constitutive of one's place in the world. It is more apparent elsewhere that Nietzsche thinks the self only gains an identity as a distinct entity by distancing itself from others. And this distancing itself is a mode of production involving measurement. By examining this aspect of Nietzsche's thesis on the self–other relation we can better locate the nature of the debt incurred within it and the conditions under which it might be repaid.

The relation between self and other is governed by will to power: by language as an expression of power, by the use of concepts to measure, interpret, draw distinctions. And according to Nietzsche, if we eliminate concepts which we impose, such as number, thing, activity and motion, then:

> no things remain but only dynamic quanta, in relation of tension to all other dynamic quanta: their essence lies in their relation to all other quanta, in their 'effect' upon the same. The will to power not a being, not a becoming, but a *pathos* – the most elemental fact from which a becoming and effecting first emerge.
>
> (Nietzsche 1967: 339)

To say that will to power is pathos refers us to the distinction between ethos and pathos which Nietzsche evokes elsewhere (Nietzsche 1974: 252). Ethos is usually understood as a way of life, one's habits and character, whereas pathos is how one is passively affected. While contractarian ethics relies on assuming that one's way of life is a given as an enduring ethos, our life, Nietzsche argues, is really pathos, a dynamic process of changing experience. The will to power is pathos: it is the movement by which experience is constituted and entities come into being such that they are in relation, can be affected and can affect.[5]

Will to power as interpretation operates within intersubjective relations where, as Nietzsche claims in reference to love, 'our pleasure in ourselves tries to maintain itself by again and again changing something new *into ourselves*' (Nietzsche 1974: 88). Measuring the other is a way of enhancing our own form, capacities and effects. But again, neither the self nor the other (whether the other is another person or a 'thing') exists in essence apart from this relation, that is, apart from 'the effect it produces and that which it resists' (Nietzsche 1967: 337). In other words, individuals, and the differences between them, are not given. They are an effect of:

> creation and imposition of forms . . . [within] a ruling structure which *lives*, in which parts and functions are delimited and co-ordinated, in which nothing whatever finds a place that has not been first assigned a 'meaning' in relation to a whole.
>
> (Nietzsche 1969: 86–7)

Will to power is this process of the constitution of place, of delimiting one from another, through the assignment of 'meaning' to effects and their interrelations. So any difference between parties to a 'contract' is an effect of will to power as productive interpretation by which entities are constituted in relation. The distance/difference between self and other is predicated upon the proximity of measurement: the credit of difference incurs a debt to the other.

If the relation of measurement between creditor and debtor is one of mutual exchange, which Nietzsche suggests in his genealogy and which contractarian ethics assumes, then it is a relation which already implies sameness. Nietzsche often notes that justice, as the fair settling of disputes (the possibility of mutual exchange without any loss of self), assumes the parties involved are already of 'approximately equal power' (Nietzsche 1984: 64; 1969: 70). At one level 'equal power' means that both parties have the power to enforce their own evaluations. At a more fundamental level 'equal power' means a balance in the distribution of productive power. The possibility of justice, that mutual understanding necessary for the repayment of debts, assumes that the selves involved are already constituted by the same mode of evaluation. That is, justice assumes that will to power as interpretation operates uniformly to produce all bodies as the same. As Nietzsche puts it in *Beyond Good and Evil*:

> To refrain from mutual injury, mutual violence, mutual exploitation, to equate one's own will with that of another: this may in a certain rough sense become good manners between individuals if the conditions for it are present (namely if their strength and value standards are in fact similar and they both belong to *one* body).
>
> (Nietzsche 1972: 174)

Belonging to one social body within which it is possible to settle one's debt to the other assumes a shared mode of evaluation by which the bodily self is constituted.

But the possibility of mutual understanding is at best limited on Nietzsche's model of self-fabrication. A social body may share a language, a mode of interpretation and evaluation, a mode of self-creation. But self evaluation occurs in relation to another and there is always a disjunction between how one evaluates oneself and how one is evaluated by another. Interpretation of the other is a translation which is a 'form of conquest' (Nietzsche 1974: 137) and reduces the tempo of the other's style (Nietzsche 1972: 41). The style projected becomes overlaid by other masks constituted through misunderstanding. The constitution of identity is dissimulation where one's absolute identity is deferred:

> Every profound spirit needs a mask: more, around every profound spirit a mask is continually growing thanks to the constantly false, that is to say *shallow* interpretation of every word he speaks, every step he takes, every sign of life he gives.
>
> (ibid.: 51)[6]

Further, while one's identity is a self-fabrication of the body using concepts one inherits, there is always a disjunction between the social concepts we share and how each person applies them:

> Ultimately, the individual derives the value of his acts from himself; because he has to interpret in a quite individual way even the words he has inherited. His interpretation of a formula at least is personal, even if he does not create a formula: as an interpreter he is still creative.
>
> (Nietzsche 1967: 403)

Finally, a crucial point which Nietzsche characteristically neglects to mention, some modes of embodiment, those specific to women, may not even be recognised as measurable within the values of the social body.

What Nietzsche exposes in his genealogy of justice and the creditor/debtor relation is that the exchange of equivalences already assumes sameness. Moreover, insofar as the parties involved are only at best *approximately* the same then evaluation involves some subtraction from the other to the benefit of the self. Social exchange does not begin with a contract between independent individuals (Nietzsche 1969: 86). It is always a matter of will to power as self-constitution and insofar as this exchange is 'successful' it assumes and promotes sameness. Yet in assuming that the other is the same, one reduces the other to the self and 'deliberately and recklessly brush[es] the dust of the wings of the butterfly that is called moment' (Nietzsche 1974: 137), that divided contradictory moment which is the site of self-creativity.

So, for Nietzsche, while the self is a socially constituted corporeal effect of one's relations with others, one's place in the world can never be reduced to another's. Despite indications of a disjunction between self and other, constituted through the relation, discourses of modernity assume sameness and encourage the desire to stay in one place. Law (which embodies notions of just and unjust) reflects a community's customs in the sense of a mode of evaluation and interpretation (Nietzsche 1984: 219; 1969: 71–6). While some law may be necessary to preserve a certain life against difference and transgression, Nietzsche objects to laws (moral or secular) which universalise notions of just and unjust and therefore impose absolute values equally upon all. In this, the notion of justice 'changes' from one which explicitly assumes sameness to one which attempts to achieve sameness of outcome. Yet what is good for one another is 'a question of who *he* is and who the *other* is' (a question of identity as measurement), and as this question cannot be answered (identity is dissimulation) then 'what is right for one *cannot* by any means be right for another' (Nietzsche 1972: 132, 139). Equating 'justice' with equal rights for all is therefore the beginning of injustice. ' "Equal rights" could all too easily change into equality of wrongdoing' because it legislates against anything rare, the possibility of difference and the need for independence (ibid.: 125). It is to this universalisation and fetishisation of value, and to its normalising effects, that Nietzsche objects.

According to Nietzsche, ' "Equality", a certain actual rendering similar of which the theory of "equal rights" is only an expression, belongs essentially to decline' (Nietzsche 1968: 91). 'Equality' belongs to 'decline' because the liberal democratic state achieves equality of outcome only insofar as it creates and 'captures' individuals, normalises them and makes them useful (Nietzsche 1978: 48–51; 1967: 382–3). What this operation of will to power creates, by enhancing certain capacities and effects at the expense of others, is what Nietzsche calls 'inverse cripples': individuals who are fragments, having 'too little of everything and too much of one thing' (Nietzsche 1978: 138). To function at all such partial individuals must be part of a larger system: equality of rights does not bring the freedom it promises but produces dependence through normalisation.

Relating Nietzsche's notion of will to power as the productive measurement involved in self-constitution to his claim that equality is only possible if equality is already actual, suggests that democratic institutions do not even achieve equality of outcome. He says as much when claiming that the democratic, 'moral' individual constitutes its ethos by negating the value of the other's difference:

> Slave morality says No to what is 'outside', what is 'different' what is 'not itself'; and *this* No is its creative deed. This inversion of the value positing eye – this *need* to direct one's view outward instead of back to oneself – is the essence of *ressentiment*; in order to exist, slave morality always needs a hostile external world; it needs, physiologically speaking, external stimuli in order to act at all – its action is fundamentally reaction.
>
> (Nietzsche 1969: 36–7)

The consistent self, the one who can enter contracts and make promises, is produced and maintained through the operation of will to power as evaluation, by exploitation, appropriation, through the imposition of a particular form and through the exclusion of others. The self is constituted as singular, unified and proper by this negation of the other which is supported by a system of conceptual oppositions (Nietzsche 1972: 21). These 'antithetical values', to use Nietzsche's terminology, remove any ambiguity in the self–other relation and establish a particular kind of self as the norm by marking off and devaluing its opposite.

Now Nietzsche, like Foucault, would have us believe that his overman is not guilty of this misappropriation of the other: creative self-fabrication is built not upon negation of the other's difference but upon a mode of self-affirmation which 'seeks its opposite only so as to affirm itself more gratefully and triumphantly' (Nietzsche 1969: 37). But, ironically, even the social body of equal and harmonious forces, which Nietzsche evokes as a sign of true justice, exists as such by marking itself off from an 'outside' to which it is hostile:

> Even that body within which, as it was previously assumed, individuals treat one another as equals – this happens in every healthy aristocracy – must, if it is a living and not a decaying body, itself do all that to other bodies which the individuals within it refrain from doing to one another: it will have to be will to power incarnate, it will want to grow, expand, draw to itself, gain ascendancy – not out of any morality or immorality, but because it *lives*, and because life *is* will to power.
>
> (Nietzsche 1972: 175)

Some commentators (e.g. Warren 1985: 202–5) point to such statements as evidence of Nietzsche's tendency to illegitimately apply his ontological doctrine of will to power to justify the necessity of political domination. However, the reading of will to power I have provided suggests another interpretation: even within the pretence of equality, whether within a 'healthy' aristocracy or a nihilistic democracy, the self, or the complex of selves rendered equal, maintains itself by marginalising others deemed inappropriate to the system. Nietzsche is not necessarily justifying political domination. He is exposing the possibility that even a political system which claims not to exercise domination, and which claims equality of outcome, is merely a disguised and nihilistic mode of domination.

If there is a difference between a 'healthy' and 'unhealthy' relation to the other, between an ethics based on an aesthetics of self and one based on normalisation, it is that creative self-fabrication, rather than negating the other's difference by reducing the other to the self, constitutes a distance, as difference, between self and other. But, significantly, there is no escaping a debt to the other when making the moment of self-fabrication one's own: the pathos of distance *within* the self, necessary for a creative re-constitution of self, is, as with democratic normalisation, predicated upon a certain relation to the other. According to Nietzsche, in the passage defining the pathos of distance within the self referred to above:

> Without the *pathos of distance* such as develops from the incarnate differences in classes, from the ruling caste's constant looking out and looking down on subjects and instruments and from its equally constant exercise of obedience and command, its holding down and holding at a distance, that other, more mysterious pathos could not have developed either, that longing for an ever increasing widening of distance within the soul itself.
>
> (Nietzsche 1972: 173)

This distancing from the other, necessary to an aesthetics of self, has its productive effects: creative self-formation, no less than democratic normalisation, incurs a debt to the other. This applies not only to relations between classes but also to relations between the sexes. To the extent that Nietzsche excludes women from the possibility of self-overcoming, he effectively builds his personal aesthetics upon the bodies of women. However, this does not have to be the case: it all depends on acknowledging that the social value attributed to the artistic self is generated through the other's generosity. It is to the operation of distance between the sexes, its effects on women and the possibility of women's artistry that I will now turn.

WOMEN'S BODIES: BEYOND VIRTUE AND SHAME

Just as measurement is involved in the constitution of any self separate from any other, Nietzsche suggests that men create an image of woman in order to shore up something about themselves (Nietzsche 1974: 126). In particular, the man fit for contracts, who conforms to an unchanging image of himself, requires a certain construction of the other to affirm and maintain the appearance of self-consistency and autonomy. This reactive approach to the other does not have to be explicitly denigrating. A man can maintain himself by constructing an ideal and essential image of woman which is simply complementary to himself, yet designed for his consumption. This image still serves to affirm the self as unchanging: it silences the 'noise' of other possibilities, the noise of the 'forgotten'. As Nietzsche puts it in *The Gay Science*:

> When a man stands in the midst of his own noise, in the midst of his own surf of plans and projects, then he is apt also to see quiet, magical beings gliding past him and to long for their happiness and seclusion: *women*. He almost thinks his better self dwells there among the women.
>
> (ibid.: 124)

The truth of woman, the eternal feminine, promises to affirm an unchanging self. But as identity is constituted in relation, the self which posits itself as autonomous and transcendental is not complete without incorporation or negation of what is other: man's desire is to possess this image of woman which he has constituted in relation to himself.[7]

To those who seek possession, Nietzsche issues a warning:

> [man thinks] that in these quiet regions even the loudest surf turns into deathly quiet, and life itself is a dream about life. Yet! Yet! Noble enthusiast, even on the most beautiful sailboat there is noise, and unfortunately much small and petty noise. The most magical and powerful effect of woman is, in philosophical language, action at a distance, *actio in distans*; but this requires first of all and above all – *distance*.
>
> (ibid.)

Possessing the image of woman as other to the self does not bring the omnipotence or self-completion promised. If woman were the complementary image man constructs, possessing this image would bring a kind of death to the self. It would efface the distance within the self necessary for the incorporation of experience in self-overcoming. While conformity relies on constituting and possessing an image of woman, under the pretence of autonomy, creative self-fabrication relies on maintaining a distance from this image. An aesthetics of self requires sexual difference: a 'noble' mode of valuation, a spontaneous mode of self-affirmation which does not seek to negate its opposite (Nietzsche 1969: 37).

But in distancing himself from woman, the creative man still incurs a debt to her. In the definition of active self-evaluation just given, Nietzsche implies an original distance between self and other. Yet, as I have suggested, he also acknowledges that even in creative self-fabrication the 'pathos of distance' involved is located at 'the origin of language itself as an expression of power' where the 'noble' spirit names itself, gives itself identity and value 'in contradistinction to all the low, low-minded, common and plebian' (Nietzsche 1969: 26). The distancing/differencing effected by will to power in self-overcoming materially constitutes woman as other to the aesthetic self. While the key to creativity lies in maintaining this action at a distance, something remains to be said about its effect on women.

Nietzsche not only claims that the creative man must distance himself from the image of woman he necessarily constitutes, he also claims that 'woman forms herself according to this image' (Nietzsche 1974: 126). This suggests that women are only artistic insofar as they are actors of a role imposed upon them. For women to be artistic in the proper sense would require the ability to incorporate experience according to one's own plan. This requires distance within the self between the present self and the concept or image towards which one aspires, which in turn is predicated upon a distance between self and other.

In the extract given above from *The Gay Science*, there are two modes of self-constitution apparently open to women in relation to men: proximity, resulting from possession by a man, and action at a distance. The first, from a woman's perspective, requires her unconditional submission to the concept of unfathomable depth man has of her. In obeying man in this way, women think, according to Nietzsche, that they will find 'depth for their surface' (Nietzsche 1978: 67). But, in submitting to men's needs, women reduce the distance between themselves and the other and hence the distance within themselves necessary for the creative incorporation of experience. Nor do they find depth for their surface. Like the actor they reflect forms not their own, merely repeating themselves according to an image provided by others.

Submission results in the constitution of woman's bodily self as a calcified image of shame. Calcified because submission collapses the difference between her appearance (surface) and the concept of unfathomable depth man has of her (Nietzsche 1974: 125). Such a woman *is* the concept, the truth of woman, fetishised. Submission brings shame in two senses. It involves being sexually possessed by a man and, connected to this, is the shame involved in the revelation through submission that woman is not the profound, unfathomable depth, the mysterious eternally feminine, which man's desire seeks. In submitting to man's desire, in giving up everything that she could be, woman's shame is constituted in revealing herself as surface. The shame deals a double blow when man loses interest. Again, to quote Nietzsche:

> There are noble women who are afflicted with a certain poverty of spirit, and they know no better way to *express* their deepest devotion than to offer their virtue and shame. They know nothing higher. Often this present is accepted without establishing as profound an obligation as the donors had assumed. A very melancholy story!
>
> (ibid.: 125)

The second mode of self-constitution Nietzsche attributes to women is action at a distance. From a woman's point of view this involves maintaining one's virtue where virtue means both distance from man's desire as well as maintaining one's difference. This woman maintains the appearance of being unfathomable and changing above the shame of being surface. Or, as Nietzsche puts it:

> old women are more sceptical in their most secret heart of hearts than any man: they consider the superficiality of existence its essence, and all virtue and profundity is to them merely a veil over this 'truth', a very welcome veil over a pudendum – in other words, a matter of decency and shame, and no more than that.
>
> (Nietzsche 1974: 125)

Action at a distance means maintaining the concept of unfathomable changeability – this is woman's virtue. But there is a catch: man's desire, whether he is artistic or democratic, is maintained. Also, the sexual 'difference' so constituted is in accordance with a concept given by man. It is in man's interest, rather than woman's, that this distance, as 'difference' in terms of man, is maintained.

For a start action at a distance, in 'philosophical language' (as Nietzsche stresses) does not bring autonomy. Action at a distance is defined philosophically (in the language of Newtonian physics) as the idea that one body can affect another without any intervening mechanical link between them. The bodies are separated by empty space yet when one moves so does the other. Woman is still moved by man's desire: a kind of mimicry is implied where woman is changeable only to the extent that man's interpretations move her. This 'action at a distance' does not distance woman from the other, nor does it allow the distance within herself necessary for her self-overcoming. In fact the mimicry implied in woman's virtue of unfathomable changeability is similar to Dionysian experience described by Nietzsche in the *Twilight of the Idols*. Here:

> the entire emotional system is alerted and intensified: so that it

> discharges all its powers of representation, imitation, transfiguration, transmutation, every kind of mimicry and play-acting, conjointly. The essential thing remains the facility of metamorphosis, the incapacity *not* to react (– in a similar way to certain types of hysteric, who also assume any role at the slightest instigation). . . . [The Dionysian individual] enters into every skin, into every emotion; he is continually transforming himself.
>
> (Nietzsche 1968: 73)

This kind of changeability is creative and Nietzsche explicitly ties it to a feminine disposition of dissatisfaction and histrionics (Nietzsche 1974: 98–9, 317). But it is only a precondition to change. To be productive the immediacy of mimicry must be offset by the distancing within the self necessary to stage and overcome the self. This distancing is the effect of the Apollonian world of images and language, that is, the will to power as interpretation, where the self is constituted as separate from another. But, as I have argued, what woman becomes through this action at a distance is in accordance with a concept provided by man. So, neither in submission to the democratic man nor at a distance from the artist do women embody the kind of aesthetics of self enjoyed by Nietzsche's overman. Contrary to the assumptions of some postmodern aesthetics, it would seem that man's desire to create himself anew is satisfied only if woman remains static, without a place of her own.

Nietzsche is not insensitive to the difficulties faced by woman as the object of man's desire. The imperative placed on women by men is to hold together a contradictory image of both virtue and shame, distance and submission, changeability and calcification. He claims that the comedy of love (Nietzsche 1974: 125–6) and the impossibility of harmonious relations between the sexes (Nietzsche 1969: 267) is based on the contradictory nature of man's self-constitution: the requirement of both distance and proximity in relation to the other. He also suggests that woman's scepticism about her role in relation to man and in the assumption of an essential self is founded on the impossibility of being the contradictory double image of virtue and shame which man requires. On the effect on women of this requirement Nietzsche observes:

> Thus the psychic knot has been tied that may have no equal. Even the compassionate curiosity of the wisest student of humanity is inadequate for guessing how this or that woman manages to accommodate herself to this solution of the riddle, and to the riddle of a solution, and what dreadful, far-reaching suspicions must stir in her poor unhinged soul – and how the ultimate philosophy and skepsis of woman casts anchor at this point!

> Afterward, the same deep silence as before. Often a silence directed at herself, too. She closes her eyes to herself.
>
> (Nietzsche 1974: 128)

Woman's solution to the riddle of a femininity constructed by man is to 'close her eyes to herself'. This closing is an opening in its suggestion of other possibilities for self-formation aside from conforming to an impossible image of the feminine posited by men. Man's dependence upon women conforming to an image of the feminine, as well as other possibilities for women, is suggested by Nietzsche in the following passage:

> Would a woman be able to hold us (or, as they say, 'enthral' us) if we did not consider it quite possible that under certain circumstances she could wield a dagger (any kind of dagger) *against* us? Or against herself – which in certain cases would be a crueler revenge.
>
> (Nietzsche 1974: 126)

As man's self-image depends upon woman conforming (whether in submission or at a distance) to an image that man has constituted for himself, then if woman does not conform to this image she effectively wields a dagger against his notion of self. That women can wield the dagger suggests the possibility of non-conformity, the possibility of artistry.

There are several modes of revenge open to women, several ways of distancing themselves from the concept 'woman' and re-creating the self differently. One possibility that Nietzsche mentions, in the context of woman closing her eyes to herself, is that she can find 'atonement' for her honour through bearing children (Nietzsche 1974: 128–9; 1978: 66; 1969: 267). However, as Alison Ainley (1988) suggests, Nietzsche tends to place a lower value on pregnancy in women than he does on the 'spiritual' pregnancy of the overman. A second mode of revenge is feminism of equality. But as my discussion of Nietzsche's views on equality above indicates, he does not approve of this option. While I think this opposition to feminism of equality is ultimately untenable, his most acceptable line of argument is that 'equality' amounts to turning women into men and is therefore not a distancing at all.[8]

The possibility of woman's creativity comes uneasily from Nietzsche's uncertainty about distance. In submission or at a distance, woman is not what she promises to be or what man thinks she is ('even on the most beautiful sailboat there is a noise'). The metaphor of noise suggests that women exceed the concept 'woman' which both the democratic and the aesthetic man posits. That Nietzsche's 'ethics' allows for women to be subjects of change rests on what he means by noise and this calls for a further reassessment of the notion of 'distance' in his philosophy.

Jacques Derrida (1979: 49) suggests, in his reading of Nietzsche, that perhaps woman is distance itself. Perhaps, but this needs qualification. Woman, operating at a distance, is the complementary image or the difference man posits in constituting himself as present. But the 'empty space' between them is effected by will to power as interpretation by which borders are established and bodies constituted. Distancing, will to power as the measurement of woman, is the difference which precedes, exceeds and constitutes the distance within the self and between man and his 'other' woman. Given the necessity of this other distancing, woman cannot be possessed – she exceeds the difference or distance over which man reaches for her or, more exactly, for himself. In proximity, or when possessed, woman will be noisy – there will be excess information. A woman is more than the concept man has of her. Her truth or identity, and therefore his, is deferred, and sexual difference as distancing is always already maintained.

If the truth of woman is to work for man he must turn away from her – he can't live with this concept but he can't live without it. But not only does the creative man turn away from the truth of woman he has constituted, so does the creative woman ('she closes her eyes to herself'). Nietzsche says of truth as a woman: 'certainly she has not let herself be won' (Nietzsche 1972: 13). Women do not become this essential image, even in submission. As Nietzsche puts it:

> Reflect on the whole history of women: do they not *have* to be first of all and above all else actresses? Listen to the physicians who have hypnotized women; finally, love them – let yourself be 'hypnotized by them'! What is always the end result? That they 'put on something' even when they take off everything.
>
> Woman is so artistic.
>
> (Nietzsche 1974: 317)

Even when forming herself by submitting to the concept of 'woman' which man projects, woman is acting as something other to both this concept and to herself.[9] Again, Judith Butler is helpful on this point. When performing herself a woman is not repeating an essence determined by the concepts of woman: while re-enacting meanings already socially established, there is always a difference in repetition (Butler 1990: 140–1).

So woman's artistry lies in her power of dissimulation. And, as I have suggested in more general terms, a woman's identity is dissimulation not because the concept 'woman' misnames the body which awaits it. Rather her power of dissimulation is based on the claim that, as embodied identity is constituted in relation to the other, absolute identity is always deferred. So the uncovering of the veil which is the surface of woman reveals, not the truth of woman, nor therefore man's self-presence, but further dissimulation. This 'putting on something' even when they take off everything is not necessarily a deliberate resistance to subjection. It is a feature of intersubjective evaluation: 'around every profound spirit a mask is continually growing thanks to to the constantly false . . . interpretations' (Nietzsche 1972: 51). Man's evaluation of woman, whether active or reactive, creates the mask that is woman's socially inscribed difference *in relation to him*. But the distancing involved in the constitution of woman's difference in relation to man ensures that the distance between them cannot be effaced – something will always be 'put on' which maintains a distance or difference. Men may assume they can capture the dangerous plaything they need to discover the child in themselves (to create themselves anew). But the old woman's advice to these men is: 'You are going to women? Do not forget the whip' (Nietzsche 1978: 67).

It is one thing to conclude that 'woman' is distance (or distancing) and therefore that women do not coincide with either the surface as fetish or with the truth of woman beneath. But let us not forget that, for Nietzsche, the self is an *embodied* effect of the operation of will to power as interpretation. It is therefore another thing to suggest that the concept of woman which man forms for himself has no effect on women. Derrida, for example, following Nietzsche, appears to risk this conclusion:

> That which will not be pinned down by truth is, in truth – *feminine*. This should not however, be mistaken for a woman's feminini*ty*, for female sexuali*ty*, or for any other essentializing fetishes which might tantalize the dogmatic philosopher, the impotent artist or the inexperienced seducer who has not yet escaped his foolish hopes for capture.
> (Derrida 1979: 55)

> Because a 'woman' takes so little interest in truth, because in fact she barely even believes in it, the truth as regards her, does not concern her in the least. It rather is the 'man' who has decided to believe that this discourse on woman or truth might possibly be of any *concern* to her.
> (Derrida 1979: 63)

It is necessary to qualify Derrida's distinction between the 'feminine' and an 'essentializing fetish'. Women may not coincide with either, but the distance/difference between female sexuality (the surface that is a woman at any particular moment) and the feminine (the undecidable concept of woman) is what constitutes women – at least insofar as women are artistic. Even in 'overcoming' themselves women rely on concepts they have inherited whether or not they may interpret these differently from men or differently from each other. Women are not outside nor completely inside the feminine as the truth of woman. But the truth of woman, as elusive and as changeable as it is, is a name. And, as the discussion above on the relation between social concepts and embodiment suggests, 'what things *are called* ... gradually grows to be part of a thing and turns into its very body' (Nietzsche 1974: 121–2). Even if what things 'are' can never be decided, concepts of 'woman' have their material effects in the constitution of the bodily self that is a woman. A woman may not believe in man's discourse on her, but given the constitutive effects of this discourse on woman's difference, to imply, however carefully, that it doesn't concern her at all is a little hasty.

Nietzsche's understanding of the 'pathos of distance' not only exposes that normative discourses assume a male subject, but also that they rely on constructing woman in a certain way. Man creates an image of woman as other in order to secure his corporeal identity. At a distance woman's 'difference' is complementary and promises to affirm man's self-presence; in proximity her 'sameness' heralds the death of the self. There is no exchange between man and his creditor, woman. Rather, woman's 'gift' to man is his (impossible) self-certainty; the 'return' for her investment is a contradictory corporality – suspended between virtue and shame. Insofar as women fulfil this impossible role as man's other they uneasily embody these contradictory concepts without a place of their own. But, as I have argued, the operation of will to power is such that women's bodies also remain open to possibilities aside from those which position them under man. The embodied meaning of 'woman' is dispersed beyond virtue and shame, beyond the riddle of femininity Nietzsche tends to uphold.

If there is a limitation in Nietzsche's approach to the problematic of the constitution of place, it is in the suggestion, apparent at times in his work, that an aesthetics of self can avoid incurring a debt to the other. This assumption is amplified in some postmodern claims that we can avoid projecting otherness outward or that we can simply declare an end to self-identity and its attendant objectification of others. To deny that an aesthetics of self modifies or objectifies the other is merely a *disavowal* of the differential relations operating in the constitution of one's embodied place in the world. As I have argued, Nietzsche's concepts of will to power and the 'pathos of distance' suggest the impossibility of such an uncontaminating space. And that action at a distance, an aesthetics of self in its simplest formulation, still relies on keeping woman in her place is testimony to the dangers of an ethics of difference which claims not to 'other' the other.

Chapter 6

Biomedical ethics and lived, sexed bodies

I have been arguing that justice, as the gift of other possibilities for sexual difference beyond sexual duality, involves paying attention to the way in which sexed identity is constituted as an embodied effect of social discourses and practices which position women as other to men. I have also suggested that, in a social context where a certain kind of male body is already the norm, an aesthetics of self, no less than democratic normalisation, is built against images of the feminine, images which compromise and restrict the reconstitution of women's modes of embodiment. During that discussion and in previous chapters, I have addressed, in general terms, the assumptions underpinning ethics, that the individual identity comes prior to relations with others and that the individual's relation to their body is one of ownership and control. What follows is a more concrete refutation of those assumptions in the context of matters raised in chapter 1: biomedical ethics and the ethics of reproductive practices. Using accounts of embodiment from the tradition of existential phenomenology, I will move from an analysis of the place of the body in biomedical ethics to a discussion of the ethics of reproduction and finally to an analysis of the role biomedical science may have in the constitution of sexed bodies.

ETHICS AND THE CARE OF DAMAGED BODIES

As discussed in chapter 1, with reference to the ethics of surrogacy, the biomedical ethicist usually seeks to: (1) protect the freedom and autonomy of the individual, understood as sovereignty over one's body, against possible transgressions by the biomedical practitioner or others, and (2) protect the values of the 'common good' from 'undesirable' practices undertaken by individual users of biomedical services or by the practitioners themselves. Included in this understanding of the ethicist's task are the assumptions that freedom is original and equivalent to autonomy; that violation of another's freedom involves taking over control of their body; that control of another's body is just if based on consent; that the self is separate from their body and remains the same through changes in the body; and that the 'common good' consists of values, however vague, which are in the interests of everyone.

While the biomedical ethicist sees their task as weighing up competing claims for sovereignty over a body, there is much to suggest that the body is not a simple passive object over which the individual has real or even potential control. In saying this I am not suggesting a return to a pre-critical notion of an innate biological complex which dictates our every move. Rather, I'm suggesting that, as we are never without a body, a notion of an individual where the body is given the secondary status of an appendage subjected to property rights would seem to assume much while explaining little. On this model, biomedical practice, which deals with and brings about changes in the body of another, is perceived as a potential threat to the individual's freedom, autonomy and property rights. However, as Daniel Callahan argues (1984),[1] while this assumption of autonomy may be necessary in medical ethics to uphold patients' rights, it does tend to diminish the sense of obligation others may have towards us and vice versa. Callahan's is a concern shared by Gilligan and the feminist ethicists of care discussed in chapter 1. Taken to its logical conclusion, this notion of freedom and autonomy discourages the ideal of community and interdependence and encourages a sterile and uncaring relationship between biomedical practice and the individual. Moreover, a problem not addressed by these authors is that this idea of autonomy belies the possibility I have already argued for, that the body contributes to what we are in ways other than as a thing to be controlled. If the individual's relation to their body is beyond a causal relation to a piece of property, the ethicist's project of protecting the individual's autonomy may not only be misplaced but also detrimental.

Some anomalies may help to clarify the relation between self and body. In his entertaining book, *The Man Who Mistook his Wife for a Hat* (1985), the neurophysiologist Oliver Sacks describes cases where people think they have a limb where there is none and, conversely, where they think the limb they have belongs to someone else. There is the man who discovered someone else's leg in his bed, but found when he attempted to throw it out that he somehow came after it (Sacks 1985: 54). When asked where his own leg might be he replied, somewhat disconcerted, that he had no idea – it seemed to have disappeared and been replaced by someone else's. And conversely, and more commonly, there is the phenomenon of the phantom limb where amputees experience their bodies as including a limb which is in fact missing (ibid.: 63–6). The memory of the limb or the 'limb-image' is so real to the amputee that they often experience pain in the missing part of their body; nor can amputees use an artificial limb unless the phantom is real.

Now we could dismiss these examples of relations between an individual and his or her body as anomalies resulting from a psychological disorder which distorts the reality of the body. But as Sacks points out, these people are in all other respects 'reasonable and clear headed'. What these examples might suggest is that the individual has a particular relation to his or her body which is not accessible to the casual observer, including the biomedical practitioner and the ethicist. Illness or any radical change in the body may represent not only a shift in one's experience of one's body (as if the body were separate from the self) but a shift in oneself *per se*. Or, in the case of a phantom limb, there would seem to be a resistance to a change in the body as it is observed, by the body as it is lived. And the fact that the body as it is lived can include a prosthesis suggests that the borders of the body as it is lived do not coincide with the borders of the body as it is observed. It is not so much that the individual stands above his or her body in appreciation of its social significance and in control of its capacities but that the capacities of the body, its habits, gestures and style, make up what the self is in relation to the social and material world.

These suggestions about the relation between the self and the body can be applied to more commonplace changes in the texture of the body. Drew Leder, in his book *The Absent Body* (1990), describes the fracture in the structure of the self which can occur with the experience of pain. A man is playing tennis. He does not reflect upon his body, his racquet, his partner or the ground between them. His posture and movements are engaged with the environment without explicit thought or will. But a sudden sharp pain disrupts this immersed activity: he gives up the game, focusing instead upon the wellbeing of his body. When his body becomes the focus of his attention, he can no longer engage with the world or the other. What turns out to be a 'coronary occlusion' is followed some time later by another leading to profound changes in this man's social identity. Beyond any reduction in capabilities, he experiences a shift in his relation to his body and to others. He is more self-conscious of his body, he anticipates pain and the 'everyday concerns of others recede as he finds himself thinking more frequently of death' (Leder 1990: 18). The body as subject (dispersed and open to the world prior to the distinctions between mind and body, self and world, self and other) has become the body as object.

What Leder provides, following Merleau-Ponty, is a phenomenological account of how illness effects the self as a lived body. In the *Phenomenology of Perception* (1962), Merleau-Ponty, arguing against both realist and idealist concepts of the human body, suggests that while it is possible to think of one's body as an object in relation to other objects, it is not like any other object in the world. Unlike other objects, a person's body is always with them and not observable to them from the outside (Merleau-Ponty 1962: 107). One's body is not an objective body in objective space. Rather, as the place of one's engagement in the social and material world, the body institutes a spatial unity, 'a link between a here and a yonder', and temporal unity between 'a now and a future' (ibid.: 140). And this unity between body-subject and world is achieved through the body's directional activity.

Merleau-Ponty's description of the body's directional activity suggests much about the relation between self, body and world. He argues that the lived body is open to different possibilities within the same situation: the relation between a body and its world is ambiguous. This ambiguity is resolved into an actual possibility through the body's motility. And the body's motility constitutes a spatial and temporal unity between one's body and its world. The body's motility is governed by the task being undertaken and the corporeal attitude or schema. The corporeal schema is a set of habits, gestures and conducts formed over time in relation to others (in a manner similar to Hegel's mechanism of habit formation discussed in chapter 3). So, possible ways of having the present world are limited by our motor 'memory': a series of past movements, experiences and habits peculiar to our social history (Merleau-Ponty 1962: 140). This history, in turn, informs the nature of the project being undertaken. And the project is realised through the body's motility, by the projection of our corporeal schema into the world in a way consistent with the task at hand (ibid.: 100). Trajectories for possible activity are marked out through this activity, and as the activity is governed by the project and the specificity of one's corporeal schema, then the meaning of the perceptual field (the relative significance of objects) is determined by both the goal of the activity and the corporeal schema of the actor.

Not only is the world incorporated into one's lived body-space through the projection of a corporeal schema but the corporeal schema is modified in the process. That is, the project polarises and gathers together the body: the body, its senses and extremities are unified and certain aspects privileged depending on the task being carried out. This intentional activity is not directed by a consciousness separate from or behind the action: 'there is not a perception followed by a movement' (Merleau-Ponty 1962: 111). Intentional activity is not directed by a choice in the form of a representation or voluntary deliberation (ibid.: 435). Rather, the action is directed towards a future through projection of a corporeal schema and the future (and hence the choice) is constituted or actualised through the body's activity. So it is not the case that a consciousness projects itself into the world using the body as an instrument. Rather, the body's directional movement itself is a projection of the self towards a future: 'he is his body and his body is the potentiality of a certain world' (ibid.: 106).

In sum, according to Merleau-Ponty, the body's directional activity or 'intentional arc' unifies and polarises the body, constitutes and incorporates its world and synthesises the past, present and future.

> [T]he life of consciousness – cognitive life, the life of desire or perceptual life – is subtended by an 'intentional arc' which projects round about us our past, our future, our human setting, our physical, ideological and moral situation, or rather which results in our being situated in all these respects. It is this intentional arc which brings about the unity of the senses, of intelligence, of sensibility and motility. And it is this which 'goes limp' in illness.
>
> (Merleau-Ponty 1962: 136)

The body's intentional activity is responsible for the body's temporal and spatial unity. This body is not something *I have*, it is what *I am* and its motility is how I have a world.[2] Second, while I am aware in some sense of being engaged with the world, this engagement is not consciously directed. In everyday activities there is no distinction between consciousness and body, between self and world. Third, this lived body is the means by which moral and social norms, habits and gestures peculiar to our particular cultural and familial background inform our activity, and the means by which past activities are brought to bear upon our future (I will say more about the social constitution of one's corporeal schema in the final section of this chapter). Finally, the unified lived body, what one is in relation to the world, is what breaks down in illness.

To say that the 'intentional arc' or the corporeal schema 'goes limp' in illness suggests that illness represents a breakdown in the structure of the self. This collapse is equivalent to the objectification of the body, a return from engagement with the world to self-reflection.[3] As an effect of something going wrong with our absorbed activity, we are thrown back onto a consciousness of our body: a split is introduced between self and body and the body becomes an object for consciousness. Conscious deliberation now directs the body, making absorbed projection difficult if not impossible. While this causal relation between consciousness and body is taken as the norm in a liberal-empiricist tradition, Merleau-Ponty claims that it is a secondary, derivative and deficient mode of being-in-the-world.

This objectification of the body by the self limits one's freedom. Freedom as Merleau-Ponty understands it is not the negative freedom of liberal individualism: freedom from physical and intellectual interference by others. He rejects the assumptions which ground such a notion of freedom: individualism, the idea of an original, isolated 'I' and the concept of a person owning property in their body. On his model of the self, freedom is the freedom inherent in the lived body's ability to structure its world and to realise the potentialities informed by its social history. While a kind of 'positive freedom', Merleau-Ponty's is not exactly the same as that understood as 'dispositional autonomy'. For Robert Young dispositional autonomy is realised if one's particular choices are ordered by a unified life plan which expresses one's own will (Young 1986: 8–9). Not only should this life-plan be the product of one's own will, but it should remain free from 'self defeating conflicts' which may arise from particular decisions inconsistent with the plan. While dispositional autonomy can be compromised by permanent physical disabilities, poverty and emotional and social factors which restrict choices for the realisation of one's plan (Dodds and Jones 1989: 2), on Merleau-Ponty's model you can be a quadriplegic and poor and be just as free as a wealthy intellectual. Insofar as one has a 'life plan' this is not decided by the will ahead of engaged activity. Rather it is built up through one's corporeal schema which is realised through concrete projects. Nor is my freedom dependent upon the number of choices open to me or on the economic, social and physical means of realising a choice. Prior to such considerations is the condition that I not be reduced to a thing either by self-objectification or by another's interpretation (Merleau-Ponty 1962: 435) (although absolute reduction of the self to a thing is ultimately impossible for reasons I will get to).

Freedom, for Merleau-Ponty, is the condition of being-in-the world in the mode of a pre-selfconscious, pre-individual engagement with an open ambiguous situation which we structure and resolve according to a corporeal style or schema built from our particular social history (Merleau-Ponty 1962: 455–6). Freedom is this being open to a world which we are. Yet our freedom is not absolute. It 'shrinks without disappearing altogether in direct proportion to the lessening of the *tolerance* allowed by the bodily and institutional data of our lives' (ibid.: 454). At one level Merleau-Ponty is suggesting that our freedom is always limited by the style of our lived body, the capacities and conducts determined by our social or 'institutional' background which may not 'tolerate' certain projects. Also implied in his account is that others restrict our freedom, not because they interfere with an original autonomy (insofar as the self is open to a world with others, there is no original autonomy in this sense), but by an intolerance of our bodily comportment towards the world, an intolerance which may reduce us from an open-ended mode of belonging to a world to 'being in the world in the way of a thing' (ibid.: 456). Not only can others reduce our freedom by objectifying our embodied mode of being and so change this to self-reflection, but, as I've said, so can illness.

Leder's tennis player, to return to the example at hand, is open to the world and engaged with it. His body is absent, in the sense of being engaged in the world without his explicit awareness or control. Or, as Leder puts it: 'The game is made possible only by this bodily self-concealment' (Leder 1990: 71). Yet his body is also present – the tennis player is his body, his body is the fabric of his being-in-the-world. With the onset of pain there is a tear in this fabric. This 'affective call', as Leder terms it (ibid.: 73), fundamentally alters the player's mode of being-in-the-world introducing an intolerance to the project. The structure of the self is disrupted when part of the body becomes an object of attention such that normal functioning becomes impossible. It is not just the pain in-itself which spoils the game: pain introduces a split between the self and the body, the self and the world, the self and the other. Rather than dwelling with a body in a world with others, one's body, the world and the other are placed at a distance. A breakdown in the operation of the body amounts to a collapse in the structure of the self. As this collapse limits the self's capacity to act, for the structural reasons given, it can be said to constrain the individual's 'positive' freedom.

There are at least three ways in which Merleau-Ponty's phenomenological account of the body-subject challenges the concept of the individual usually assumed in biomedical ethics. First, the body is not so much an appendage to the self as the very fabric of the self. Second, as the lived body is the locus of one's being-in-the-world, one's social identity is derived from one's specific corporeal history and will vary in each case. The lived body expresses what the phenomenological tradition refers to as difference in identity. Finally, the distinction between the individual agent and the body over which she is sovereign, the distinction upon which biomedical ethics is based, is a secondary and deficient mode of being-in-the-world. Self-conscious intentionality, sovereignty over one's body, is a mode of existence which arises after a call back to the body from habitual engagement. Moreover, in this state of 'having' rather than 'being' a body, one is alienated from the world, from others and from the self. In pain, and, it would seem, in biomedical ethics, we are alone with a body which is a stranger. Yet if we throw the stranger out of bed we are sure to follow after it.

David Schenck (1986), follows the same phenomenological tradition as Leder, but takes Merleau-Ponty's account of the embodied self as a foundation for medical ethics. The body for Schenck is not just a centre of activity by which we are engaged with our environment, but as such, is *'literally our selves expressed'* (Schenck 1986: 46). The body we are with is a kind of sign out there in the world inviting interpretation by and response from others. And it is this text of our being-in-the-world which is the object of biomedical practice.

According to Merleau-Ponty, insofar as others mean something to us and we mean something to them, this meaning is conveyed through the lived body. The body, its spatial and temporal unity, its habits, gestures and conducts, expresses our existence. However, it is not the case that the lived body is separate from our existence and acts as a sign for it. Merleau-Ponty insists that the lived body is not an instrument for us to actualise our thoughts or choices, nor an expression of a meaning which comes before it. Hence the body is not a sign of consciousness, as it is for Hegel. Rather the corporeal schema embodies its signification, it expresses itself in relation to its context. Merleau-Ponty, following Sartre, uses the example of anger to illustrate his point. When I am faced with someone who manifests a threatening gesture, I do not need to infer the thought of anger behind the gesture. I read anger in the gesture itself. 'The meaning of the gesture does not lie behind it' (Merleau-Ponty 1962: 186). We cannot assume a distinction between the meaning of our existence (our social identity) and its expression because we mean nothing apart from our modes of existence and 'the body can symbolize existence because the body realizes it and is its actuality' (ibid.: 164).

We 'understand' another if we understand the meaning of their gestures and this is only because we share a common social world. Through this social world we develop habits, modes of movement and gestures which have a common meaning. So another's gesture only means something to us if it is already familiar such that we can identify with it. Or, as Merleau-Ponty puts it, '[t]he communication and comprehension of gestures comes about through the reciprocity of my intentions and the gestures of others' (ibid.: 185). Further, while we may share a common world, we bring to our relations with others a corporeal style unique to our particular history. Finally, while illness consists in a breakdown in the structure of our 'social expressiveness', we still carry that structure with us (our corporeal schema can become intentional at any time) so that we 'never become quite a thing in the world' (ibid.: 165).

To return to Schenck's point: the object of medicine is not an a-social, passive body which merely houses the self. Rather, the object of medical practice is the self expressed. And, as medicine deals with what Schenck calls the 'brokenness of bodies', it is dealing with the collapse of one's social expressiveness (Schenck 1986: 51). It is the fact that the body is the self expressed which gives medical practice its ethical dimension. A broken body invites care by others and in this call for care one gives over to medical practitioners the responsibility for restoring the social identity which is the body-self.

Merleau-Ponty's account of the body as the 'self expressed' has further implications for biomedical ethics which Schenck does not mention. The phenomenological account of the relation between the individual and his or her body suggests that, if we assume that the body is merely an appendage to the self over which we ideally have sovereignty, then biomedical ethics gives biomedical practice an unethical foundation. In the case of the so-called 'broken body', if medicine assumes that this body is merely a thing which requires fixing, a malfunctioning object of secondary concern to the self, then it is likely to perpetuate the alienation inherent in the collapse of the lived body. Iris Young's understanding of alienation is useful here. Young, following the phenomenological tradition, defines alienation as 'the objectification or appropriation by one subject of another subject's body, action, or product of action, such that she or he does not recognise that objectification as having origins in her or his experience' (Young 1990a: 168). Alienation is not about losing ownership of one's body. Rather it is about reduction of this lived body to a thing. The body-self, already objectified in its 'brokenness' may be doubly objectified by a medical practice which treats it as an object, universal in essence and a mere appendage to the self. In this, the body, and therefore the self, will remain a stranger. And biomedical ethics is unethical insofar as it contributes to a fragmentation of one's being-in-the-world in viewing its task as weighing up competing claims for sovereignty over a body.

At a time when patient consent is for the most part mandatory in western democracies, what is at stake in medical intervention and biomedical practice is not so much the autonomy of the individual (understood as negative freedom, sovereignty over one's body) as the restoration of one's embodied being-in-the-world (one's positive freedom). Further, while recovery involves moving from 'having' a body to 'being' a body, the self who emerges from this process will not be the same as the self prior to 'breakdown'. On Merleau-Ponty's model of the lived body, recovery from injury or illness is a metamorphosis, a rehabitation which while reclaiming forgotten gestures is also futural or open to other possibilities (Merleau-Ponty 1962: 164–5). Also, what is at stake in biomedical practice is not the integrity of a passive object which is fundamentally universal in its nature. As the body is the texture of the self's social identity which varies in each case, what is at stake is the identity of one's difference. The phenomenological model not only reinstates the dignity of the patient by stressing that the fabric upon which biomedicine works is the self, but it also highlights the specificity of that person's condition, however common that condition may appear to be. Finally, in claiming a difference in the identity of one's lived body, the model challenges the assumption of a common good (a set of interests applicable to any body) without abandoning that of concern to Gilligan and Callahan: the notion of community and its attendant obligation of care.

While phenomenology's notion of the lived body goes some way to challenge the foundations of biomedical ethics, there are at least two problems with the sharp distinction both Leder and Schenck make between the open lived body (the body's self-concealment in unimpeded activity) and the broken body (the disruption to the texture of the self which occurs when the body's integrity is altered). First, not all changes to the texture of the lived body can or should be understood in terms of an inhibition of (positive) freedom nor therefore in terms of self-estrangement. Not all the bodies of biomedical practice are 'broken'. The pregnant body, for example, is a common object for both biomedical practice and biomedical ethics and pregnancy obviously involves profound changes in the lived body. Yet pregnancy can hardly be thought of in terms of a collapse in the structure of the self. Schenck's foundation for medical ethics is just as impotent as the contractarian model when it comes to dealing with unbroken, sexed bodies. I will deal with this inadequacy in the following section.

A second problem with assuming that biomedicine works only on broken bodies is that biomedical ethics avoids the question of how one's 'social expressiveness' or corporeal schema is constituted before it is 'broken'. This question, which I will address in the third section of this chapter, raises the issue of the social constitution of differences in the corporeal schemas of women and men, the forms of injustice involved and the possible complicity of biomedical science in this process of production.

PHENOMENOLOGY OF PREGNANCY: RETHINKING THE ETHICS OF REPRODUCTION

Just as Merleau-Ponty's model of the lived body challenges the foundations of biomedical ethics, it also provides grounds for rethinking the ethics of reproductive practices. Besides abortion, which I will not discuss here,[4] the two kinds of reproductive practice of most concern to some feminists are surrogacy and the use of reproductive technologies. On the one hand, as I have already discussed in chapter 1, the model of autonomy as sovereignty over one's body and the contract model of social relations allows both on the basis that a woman has a right to do with her body what she will. Or, using the same assumptions, if the sex of the body is acknowledged, the practices may be deemed unethical on the grounds that the body involved is a woman's (implying that only the male body can be a legitimate subject of exchange in the market place). Or, again using the same model, the pregnant body may be said to be two bodies and surrogacy, for example, may be disallowed on the basis that it amounts to baby selling which would be harmful to the 'common good'.

On the other hand, as I have also discussed, there have been various feminist challenges to this approach to the ethics of reproductive practices. A central feminist claim is that, as the 'common good' is patriarchal, its values tend to endorse male control of women's bodies: the 'sexual contract' precedes the 'social contract', as Pateman puts it (Pateman 1988). With this consideration, the feminist ethicist assesses the ethics of reproductive practices by monitoring the extent to which they perpetuate male control over women's bodies. On this basis, surrogacy and the use of reproductive technologies are discouraged while abortion, if freely chosen, is endorsed.

A second feminist objection to both surrogacy and reproductive technologies, mentioned in chapter 1, is based on the idea that a woman's decision to become the subject of either practice is only ethical if it is autonomous and it is only autonomous if the woman is fully informed of the consequences of her decision for her future wellbeing. The notion of autonomy assumed here involves a kind of dispositional autonomy. A woman's decision to enter a surrogacy agreement or to use reproductive technology is autonomous if it does not conflict with her interests or jeopardise the ongoing realisation of her self-designated life plan. Susan Dodds and Karen Jones argue that a woman's decision to enter a surrogacy agreement (formal or informal) is not autonomous in this sense because it leads to numerous unpredictable traumas and conflicts (giving up the child and the 'usual risks' of pregnancy) (Dodds and Jones 1989).[5] Diana Kirby (in NBCC 1990: 17) makes a similar claim about the use of reproductive technologies: the autonomy of the decision to use such services is compromised by the fact that it is impossible to predict the alienation which will be experienced in the process.

What is implied in this second kind of objection is that, as the woman making the decision would not reasonably make it in the affirmative if she knew what she was in for, there must be some coercion involved even where the project is agreed to or actively sought. It is said, for example, that, in the case of surrogacy, economic need arising from economic inequalities in patriarchal society may unduly influence the surrogate's decision by overshadowing considerations of her future wellbeing. But as economic need underscores the work projects of most women in capitalist economies, and as these may in time result in various unpredicted emotional and physical traumas, this is hardly a form of 'coercion' unique to surrogacy. It is also claimed, in the case of both surrogacy and the use of reproductive technologies, that the autonomy of women involved is jeopardised by the imperative to procreate. There is little doubt that such pressure exists, but in a patriarchal society it exists for all women to varying degrees. Again the 'coercion' here is not unique to these particular reproductive practices.

This brings me back to the first objection: that surrogacy and the use of reproductive technologies increases male control over women's bodies. The problem with this objection is that, if patriarchy is understood as legitimating male control over women's bodies, particularly through regulative mechanisms to do with reproduction, then it is difficult to see why this charge is directed against surrogacy and reproductive technology rather than procreation in general. While it is necessary to acknowledge that the social context is patriarchal, and to challenge the notion of consent accordingly, I have suggested that it is problematic to then assume, as is often the case in these debates, that a woman's freedom and autonomy is preserved if she keeps her body to herself.

This is not to say that I endorse the contractarian defence of either surrogacy or reproductive technology. As discussions in previous chapters imply, there are grounds for being suspicious of both. However, as I will argue below, it is not surrogacy itself which is unethical but the contract which governs the project. And it is not the use of reproductive technology which is unethical but the kind of objectification of women's bodies which occurs in the process. If we understand surrogacy to be a project involving pregnancy under the promise that the child will be given up in the end, and the use of reproductive technology as a project to bring about pregnancy, then on the model of the lived body outlined in the previous section, the main factor which would compromise the freedom of the women involved (all things being equal regarding prior 'coercion') would be the extent to which objectification of the body (by the self or by others) occurs for the duration of the project. Here freedom is understood as the open embodied engagement with others in the world rather than isolated sovereignty over one's body. So freedom is reduced, rather than enhanced, by keeping one's body to oneself. A woman's freedom is jeopardised by the reduction of her lived body to a thing through various forms of intolerance including the appropriation of her experience by another.

There is little doubt that objectification by others occurs in both projects insofar as biomedical practice is involved. Unlike the cases raised in the previous section, the lived body of the surrogate or the woman using reproductive technology does not present itself to the biomedical practitioner as a broken body, yet it may be treated as one. However, as I argued in the previous section, insofar as this objectification can be minimised the ethical burden is on the practitioner not the woman involved. In the case of surrogacy, the surrogate's freedom may be further compromised by the commissioning couple insofar as they treat her lived body as an object, as an instrument for the realisation of a child. This limitation on the woman's freedom is more a product of the assumption that social relations are based on the exchange of body property than anything inherent to the project of surrogacy. Finally, in both surrogacy and the use of reproductive technology, and before other considerations, the ethicist shrinks the woman's freedom insofar as she or he appropriates the experience and presumes to know ahead of the project what the woman does not: exactly what the embodied experience will be.

At this point I wish to leave aside the question of the ethics of reproductive technology in order to consider the ethics of contracts which involve the maternal body but not necessarily a biomedical practitioner. While the surrogacy contract will provide my focus, the analysis is aimed at assessing, in general terms, the viability of both the contract and the phenomenological models of the self when dealing with the pregnant body and other modes of embodiment which signify womanhood.

On the contract model of social relations, the self, understood as separate from the body, remains the same throughout minor or dramatic corporeal changes. From this premise it can be assumed that the pregnant woman remains the same throughout her pregnancy despite profound changes to her body (these changes being attributed to another body, that of the foetus, or if pertaining to the pregnant body still being assumed to be separable from the self). On the basis that the self is unchanging, the woman who signs a surrogacy contract before or at the time of conception can be held to her promise to give up the child at birth. Merleau-Ponty, by equating the self with the lived body, implies that the self changes with changes to the corporeal schema. As in Nietzsche's philosophy, the self is a dynamic process of corporeal self-fabrication. On this kind of model, the woman who signs the surrogacy contract is not the same self who is asked to give up the child.

The claim that the pregnant woman does not remain the same through her pregnancy not only invalidates the surrogacy contract but has wider implications for ethics. It is a claim therefore which requires further justification, particularly given a problem with the phenomenological model of the self mentioned towards the end of the previous section. There I suggested that the sharp distinction both Leder and Schenck make between the lived body and the broken body tends to allow the assumption that all significant changes to the lived body lead to an incapacitation which invites professional intervention. As discussed, the projection of one's corporeal schema during engaged activity may meet with 'intolerance' or resistance due either to internal limitations arising from one's own body-history or to objectification of the body arising from external factors. While this 'intolerance' throws the absorbed self back onto an awareness of the body it does not necessarily involve a breakdown in the fabric of the self. A change of habits, for example, involves a heightened awareness of one's body consistent with a significant change to one's corporeal schema without a dramatic dislocation from the world, the body or from others (Merleau-Ponty 1962: 144–6). This can also be true of acquisition of a new skill (incorporating a piece of equipment into one's body space) and a sexual encounter (incorporation of another's body).

Pregnancy, to return to the body in question, involves profound changes to bodily capacities, shape and texture with attendant shifts in the awareness of the body. Yet, as Iris Young argues, pregnancy can be better understood as an expansion in the borders of the self than as a collapse of its structure (1990a: 160–74). Insofar as phenomenologists formulate the experience of being thrown back onto an awareness of one's body in terms of disruption and alienation they miss much about how we ordinarily accommodate corporeal change.

Not only has the distinction between the body's concealment in free activity (more often called transcendence) and the broken body (immanence) persisted in the existential phenomenological tradition, but there has been a tendency to equate the capacity for transcendence with men and immanence, or the inhibition of free activity, with the bodies and sexuality of women. While, to my knowledge, Merleau-Ponty is not guilty of this, Sartre and de Beauvoir certainly are. As is well known, Sartre associates 'slime' and 'holes' (passive bodies, parts of bodies and substances which pose a threat to transcendental activity) with feminine passivity, women's sexed bodies and female sexuality (Sartre 1966: 776–83). De Beauvoir, while appreciating that women's modes of embodiment are equated with immanence as an effect of defining the male body as the norm (de Beauvoir 1972: 15), tends to perpetuate this equation. Perhaps her most notorious claim along these lines is her suggestion that, while men through their futural projects create life and cultural values, the maternal body merely repeats life (ibid.: 94–5). To this claim she adds the conclusion that whilst women remain 'closely bound to their bodies' in this way, 'like an animal', they will not be free (ibid.: 97).

Merleau-Ponty indirectly corrects this tendency to reserve the category of free transcendental existence for man's activities through his critique of Sartre's concept of freedom (Merleau-Ponty 1962: 434–56). For Merleau-Ponty, as I have suggested, freedom is always limited by the history of your corporeal schema and by the fact that projection of this schema is often met with intolerance in the forms described. No human body engaged in a project merely repeats life and no body is simply free to create itself and its world anew. This distinction between the passive repetition of life and active transcendence is also the target of Iris Young's phenomenology of pregnancy (Young 1990a).

Young argues that in pregnancy as in other modes of embodiment, the self is her body. This lived body is not the unified singularity of contract theory, nor is it either absolute transcendence or passive immanence. Against the distinction between self and other, central to contract theory, Young suggests that pregnancy is a mode of embodiment which undermines the assumption of a given border between the inside and the outside of the self. In pregnancy it is not clear where one body ends and the other begins. For the woman herself, the first movements of the foetus can produce a sense of division within the self (Young 1990a: 163). But this division is not between the self and another singular self inside the self (as contract theory would have it): the inside of the self is both other to the self and part of the self. Nor is it clear where the pregnant body ends and the world begins (ibid.). The border between the self and the outside changes with subsequent changes to modes of engaged activity. The change of border between self and world effects a change of habits, gestures and posture which, on a phenomenological model of the lived body, is equivalent to a change in the self. So for Young the self at the end of pregnancy is not the same as the self prior to pregnancy. Pregnancy is not a passive waiting for the birth of another self, but an active period of change to the woman herself (ibid.: 167).

These changes in the integrity of the body do not necessarily alienate the pregnant self from the world. While some sense of alienation is involved this is as much a product of the social significance of pregnancy and the appropriation of the experience by obstetrical medicine as it is due to anything inherent in pregnancy (Young 1990a: 168–70). Pregnancy does not precipitate a radical split between self and world in that the pregnant woman may find it difficult to negotiate the shifts in the border between herself and the world, but she negotiates them all the same. Yet, because of the resistance to projection of this corporeal schema, the pregnant self is not absolute transcendence. Against the distinction between transcendence and immanence apparent in the phenomenological tradition, Young's analysis suggests that pregnancy involves a substantial change in the integrity of the body and hence a change in self without a paralysing division between self and world, self and other, self and body. Such an aesthetic mode of existence is not peculiar to pregnancy. Rather, Young's analysis is directed against any notion of the body is a pure medium for the self's activities.

For Young, pregnancy is a mode of being-for-self where changes in the lived body bring about not only the emergence of another within the self but a change in the self as well. Insofar as contractarian ethics conceives of the pregnant body as two discrete bodies in one, or any body as atomised and self-contained, it overlooks the fundamental interdependence of self and world, self and other – an interdependence which is effected through the body. And insofar as contractarian ethics assumes that the self remains the same through significant corporeal change, and to the extent that phenomenology views a change in the body's integrity negatively as disruption and a deviation from the norm, neither can account for the flexibility of our being-in-the-world: the self-creativity inherent in corporeal change. On the other hand, if we think of pregnancy as an example of a change in the integrity of one's body which effects the production of another within an expansion of the self, then this not only refutes the general assumption that the maternal body is engaged in passive repetition of life, but it challenges the validity of contracts governing this mode of production.

As I have suggested, the validity of the surrogacy contract depends on the assumptions that the bodies of the 'surrogate' and the commissioning couple are distinct from their substantial identities; that you can arbitrarily divide the bodies of those involved in conception to determine what is exchanged by whom and therefore who owns the product of exchange; that the identity of any of these bodies is not affected by the contract; and that the identity of the surrogate remains the same throughout pregnancy. However, as I have argued in chapter 4, following Mauss's model of the gift, the social identity of each party in relation to the other is constituted through the 'contract'. And, from my reading of Nietzsche's philosophy, it can be argued that the social identity constituted in this 'exchange' is coincident with an evaluation and reconstitution of the body. Nietzsche also claims that the idea of a consistent self behind various activities and changing modes of embodiment is a moral fiction, a fiction which only approaches reality if the body is trained to be consistent through a system of punishment. However, as I suggested in chapter 5 and again here, even if this training is effective enough to produce a body which dutifully repeats a limited number of socially desirable acts, this body and hence the self will still change constantly and sometimes dramatically, through a change in habits, through an injury and, to take the case in point, through pregnancy.

If the social identity of the self cannot be distinguished from the lived body by which it is actualised and if one's self-image cannot be distinguished from the living of this body as a whole, then it should not be surprising if changes in the body effect changes in the structure and fabric of the self. Therefore, in the case of surrogacy and in the light of Young's phenomenology of pregnancy, the woman who made a promise at the time of conception is not the same self who is asked to keep it. That the self changes with a rehabitation of the lived body explains what contract theory cannot tolerate: that a surrogate mother may begin by treating her body as an object of exchange but end by wanting to keep the product of her labour. To allow this change of mind does not involve assuming that pregnant women are more fickle than everyone else. Rather it involves recognising that such a significant change of body is a change of mind. To hold a surrogate to her contract is to assume that the product of her labour is not part of herself. The same assumption, that the border of a body as it is observed is the same as the border of the body as it is lived, would have us cure amputees of their phantom limbs even though they cannot function without them.

Further, if the lived body is the fabric of the self, it is not necessary to argue about which bits of body property are exchanged at the time of conception in order to determine who owns the end result. The issue is rather which embodied self was altered irreversibly between conception and birth. What links the birth mother to the child at the time of birth is a substantial change in her corporeal schema; all that links the genetic father to the child (or the genetic mother if *in vitro* fertilisation is used) is, as Giulia Sissa suggests, an assumption about self ownership based on body-property (Sissa 1989: 133). Whatever the involvement of the commissioning couple in the surrogate's pregnancy, the project and the attendant changes in corporeal schema are the surrogate's. To claim otherwise is to appropriate the surrogate's corporeal schema, reduce her body to a thing and compromise her freedom accordingly.

While the surrogate cannot be justly held to any contract to give up her child, nothing I have said precludes her from doing so if she desires. But this would involve a different project to pregnancy. While pregnancy involves emergent incorporation of another within an expansion of the self, giving up the child at birth or shortly after would involve a further process of rehabitation in relinquishing part of the self to another. Again, Mauss's concept of the gift is useful for understanding what is at stake in such a project. As indicated by my discussion of the gift in chapter 4, if a woman gives her child to another this does not primarily involve exchange of body-property understood as separate from the self. Rather, what is given is the substance of the giver. What is constituted through this gift is the substantial identity of both the giver and the recipient in relation to the other and an enduring social bond between them which obligates the recipient to the birth mother. Giving one's child to another is not unethical but a profound expression of the gift. What is unethical is the assumption that the gift can be annulled, by the one who invites it, through any form of return. To assume that the recipient can pay out the gift is to reduce the birth mother's body to a thing and her gift to a commodity and this denies the contribution she has made through the gift to the recipient's identity.

THE SEXED BODY-FOR-OTHERS

So far in this chapter I have skirted around the claim, central to feminist opposition to some reproductive practices, that these legitimate male control over women's bodies. However, I have implied that self-control or sovereignty over one's body is not the central issue in addressing women's social subordination with the suggestion that, if the self is actualised through projection of a corporeal schema onto the world of others, your freedom is compromised rather than preserved by keeping your body to yourself. Yet it would be misleading to imply that the consequences of self-formative projection are the same for women as for men in patriarchal social relations. As the analyses of the constitution of embodied identity and sexual difference in previous chapters indicate, I think the central issue in redressing women's social subordination within patriarchal social relations is not so much male control of women's bodies as the ways in which women's bodies are socially constituted in relation to men. It is here, through the operation of 'intolerance', in both senses of the word, that any limitation on women's freedom can be located. For the remainder of this chapter I will again take up this problematic of the constitution of a sexed embodied ethos, this time through a phenomenological account of the lived body and with special reference to the possible role biomedical discourses may play in its constitution.

The first question to be addressed then is the one eschewed in Schenck's distinction between the transcendent body and the broken body: what determines the capacities and interests as well as the 'social expressiveness' of the sexed body before it is obviously broken or before it expands? Second, is there injustice to be found in this mode of production (is intolerance sex specific), and finally what role does biomedical science play here? Merleau-Ponty indirectly attends to the first question through his claim that the lived body is constituted by its dwelling in the world. That is, the capacities and habits, and therefore the interests, of any body do not arise separately from its engagement with others nor from the discourses and practices which make up the world in which it dwells.

For Merleau-Ponty, your corporeal schema is never individual: it is fundamentally intersubjective and specific to your social and familial situation. Further, as a corporeal schema is constituted in relation to others, it is ambiguous. Insofar as any body claims absolute self-identity and difference from the other, through building a partition between their body and the body of the other, this ambiguity is suppressed. It is on the basis of the socially specific mechanism of this partitioning, governed by circulating discourses about what is proper to bodies, that we can derive an account of the social constitution of sexed bodies.

Merleau-Ponty explains the intersubjective nature of the lived body, and its socially specific constitution, through an account of its early development in 'The Child's Relations with Others' (in Merleau-Ponty 1964). He argues that the child does not initially carry a distinction between the inside and outside of itself, nor therefore can the child distinguish between external perceptions (its extroceptive body as it is seen and touched by others) and introspective perceptions (the body as it is lived, feels, sees and touches). And as there is no distinction between self and world, self and others, the child is not originally conscious of itself or of others.

In this undifferentiated state of 'anonymous collectivity' as Merleau-Ponty puts it, the relation to others is one where 'the other's intentions somehow play *across* my body while my intentions play across his' (Merleau-Ponty 1964: 119). That is, as the child's motility develops, it picks up the corporeal schemas of others and thereby incorporates the gestures and habits of those within its immediate body-space. Similarly, through non-conscious experimentation with movement and insofar as its experience of its own body and that of the other are inseparable, the child's gestures are transferred to others. The distinction between self and other, and hence consciousness, arises through a gradual differentiation of the child's experience of its own body from its experience of the bodies of others. Merleau-Ponty accounts for this process through a critical appropriation of Wallon's and Lacan's models of the 'mirror stage'.

The distinction between self and other is based on recognising a difference between introspectivity and extroceptivity. This can occur through the child's recognition of itself in the 'mirror'. The child's mirror-image is at the same time itself, separate from itself and what the other might make of it. In recognising itself in the mirror, the child learns that it is an object for another and therefore distinct from the other. However, the distinction from the other cannot be absolute. According to Merleau-Ponty, a system of indistinction is established between the child's introspective body, its visual, extroceptive or objectified body and the body of the other. Insofar as the child identifies with the image of itself it cannot easily distinguish between what it lives, what the other lives and what it perceives the other is doing (Merleau-Ponty 1964: 135). This tripartite system is one of 'syncretism': the reciprocal transfer of movements and gestures between dispersed bodies.

So, the emergence of the child's own corporeal schema and sense of itself separate from others, while based on objectification by others, occurs through the organisation of its body in reciprocal relation to others. And this constitution of the body-subject through the other occurs without conscious intervention: the child assumes the other's gestures and conducts through mimesis and projects its own conducts onto the other through identification with the other, that is, through 'sympathy' and 'transitivism' (Merleau-Ponty 1964: 145, 148). That a child's corporeal schema, which is itself, is organised through mimesis and transitivism is why it can be said that one's lived body is socially constituted: it is built on the invasion of the self by the gestures of others who, by referring to other others, are already social beings (ibid.: 144). The kinds of conducts incorporated into the lived body, as well as their social significance, and hence the limits to your potential modes of being will vary depending on with whom you associate and under what circumstances. Presumably then, insofar as social discourses expect and encourage differences in male and female comportment, these differences will be incorporated into the corporeal schemas of children.

Not only is the lived body socially constituted in this way, but as it is based on a confusion between self and other, it is fundamentally ambiguous. Merleau-Ponty's version of the 'gift' consists in his understanding of this ambiguity. The lived body is only the self by virtue of being an object for others, and yet by identifying with and differentiating itself from this image of itself, which it is not, the self lives through another. So while my position or ethos is constituted through the building of a partition between the inside and outside of my body, between my body and the others, this difference is not absolute: through this very differentiation my conducts are given to, and take place in, the world of the other's body as they live from and with me. The lived body is neither exclusively a subject nor an object but both (Merleau-Ponty 1962: 167): it is constituted and lives as an interworld of potentiality opened onto others. And as this corporeal schema is prepersonal rather than a singularity (ibid.: 84) it cannot be pinned down: what it is remains ambiguous.

Some of this ambiguity is removed when the child says 'I', when it takes up one point of view as the subject of language (Merleau-Ponty 1964: 151). In this, the indistinction between self and others is diminished as is the transfer of thoughtless gifts (ibid.: 153), that unconditional giving of the self to the other inherent in the confusion between the two. Yet the structure of language is such that each person, while being an 'I' for themselves, is also a 'you' for others. As the self as a lived body remains caught between subject and object, syncretic sociability still structures an adult's embodied existence even with a lived distance between self and other (ibid.: 154).

The adult encounter with another is such that:

> [w]hether it is a question of another's body or my own, I have no means of knowing the human body other than that of living it which means taking up on my own account the drama which is being played out in it, and losing myself in it.
>
> (Merleau-Ponty 1962: 198)

I encounter others primarily through a familiar dwelling with them rather than in the mode of intellectual recognition. I can only begin to 'know' the other if we have conducts and therefore a social history in common (ibid.: 186). And even then, engaging with others is a matter of lending and borrowing gestures; the self and other are actualised in the encounter through the 'blind' recognition and identification with the conducts of another (ibid.: 185). This corporeal reconstitution through the other is the realm of the gift, the realm of syncretic sociability, of ambiguous, open-ended, incalculable embodied existence. Insofar as I distance myself from my body or the body of another and attempt to know it, knowledge of that lived relation is lost. I can only *live* my relations with others, and in this familiar dwelling with others I 'lose myself': the self-conscious self, the singular 'I' who attempts to know, is dispersed. So what the self is in relation to others cannot be calculated.

I have already noted that the ease of an encounter with another is limited by the extent to which you already have gestures in common. Faced with a stranger with a different cultural history and hence a different corporeal schema one's own lived body may exhibit intolerance or resistance to the encounter. As Jennifer Biddle (1993) argues, such resistance may manifest as dis-ease: physical discomfort or, under certain circumstances, illness.[6] In particular, if the lived bodies of men and women have a different history then it should not be surprising if some solicitude governs an encounter between them. Injustice arises when, rather than reducing such unease by giving to the other through the labour of dwelling-with, the other is made familiar by denying their ambiguity. And it is this kind of injustice which governs the constitution of women's embodied modes of existence in patriarchal social relations.

Merleau-Ponty notes two general ways in which the ambiguity of the relation to the other can be effaced: by treating the other's difference as an absolute difference or by assuming the other is identical (Merleau-Ponty 1964: 102, 106). Both involve atomistic dualism and 'intellectual rigidity': the other's identity is assumed to be fixed, natural, given and knowable apart from, and unaffected by, my embodied dwelling with them. Reducing the other's corporeal difference to absolute difference involves placing the other at a distance and attributing to them those characteristics the self does not want (Merleau-Ponty 1964: 103–4). The other is thus reduced to a negative complement of the self. Following de Beauvoir, Merleau-Ponty instances sexual difference as exemplary of this kind of 'social dichotomizing': 'men who, by virtue of the established myths . . . , do not want to be weak and sensitive and want to be self sufficient, decisive, and energetic, project on women exactly those personality traits they do not themselves want to have' (ibid.: 104). Ambiguity can also be denied by effacing the other's difference altogether. In this we

> refuse to see among men [sic] even the most striking differences of *situation* – differences which pertain to the collectivity in which they have lived and received their initial training. There is an abstract or rigid liberalism which consists in thinking that all men are *identical.*
> (ibid.: 106)

As I have argued in previous chapters, the constitution of sexual difference is such that the 'established myths' about masculininty and femininity constitute women as other to man and man is maintained as the norm as a consequence. In terms of Merleau-Ponty's model of embodied dwelling with others, this 'social dichotomizing' effaces the ambiguity or open-endedness of a woman's lived embodiment, an ambiguity inherent in any encounter with another. Then, in the name of equality and sexually neutral contracts any differences in women's embodied modes of existence are ignored by claiming that women are the same as men.

How then do 'established myths' about sexual difference affect the constitution of women's modalities of bodily existence? It is again useful to begin with Iris Young's work in this context. In her aptly titled paper, 'Throwing Like a Girl' (1990a) Young sets out to explain why (white, western, middle-class) women tend to make less use of their body's spatial and extended potentialities than men (why women tend to 'throw like a girl'). Using Merleau-Ponty's model of the constitution of the corporeal schema, Young accounts for those restrictions which signify a 'feminine' bodily comportment in terms of the way in which social attitudes about sexual difference are incorporated into women's styles of existence. The 'social dichotomizing' referred to above allows women to be treated as fragile, passive objects and reduces women's bodies to things to be looked at and acted upon. A woman's bodily comportment actualises this reduction insofar as she experiences herself as '*positioned* in space' rather than an active potentiality absorbed in activity (Young 1990a: 151). While acknowledging differences in bodily comportment between women, Young's general claim is that differences in the social situations of men and women and social attitudes about sexual difference are lived in and through the bodies of women. Or, as she so graphically puts it: 'Women in sexist society are physically handicapped. Insofar as we learn to live out our existence in accordance with the definition that patriarchal culture assigns to us, we are physically inhibited, confined, positioned, and objectified' (ibid.: 153).

One point which Young's account suggests for my analysis is that, if women's freedom is limited in patriarchal social relations this occurs as much prior to as within a particular project such as pregnancy. And the limitation is not due to male control over women's bodies or to objectification *per se* (the self is constituted through objectification of and by others). Rather, the limitations on freedom are primarily due to ways in which women's bodies are constituted and valued in relation to men and in relation to circulating discourses about sexual difference. Further, any dis-abling of women effected by this social dichotomising is reinforced by excluding women from particular projects on the basis of a different bodily comportment and/or on the basis of existing asymmetrical power relations between men and women. And the injustice is trebled if, to ease any solicitude, differences in embodied modes of being are effaced by insisting that women are or should be the same as men. The point is not that women should learn to throw like a boy. Nor is it only that styles of existence are an effect of training according to what is assumed to be proper to sexed bodies. The point is also that these styles of existence are always open to other possibilities in their ongoing actualisation. Relieving the injustice of this mode of production of sexual difference lies in questioning the discourses and practices which close off the ambiguities inherent in intersubjective embodied existence.

This brings me to the question of the role biomedical science may play in the constitution of the sexed body. While phenomenologists such as Leder and Schenck acknowledge that the lived body is constituted through its engagement with others in the world, the possible effects of scientific discourses on this engagement remain unclear. The distinction between the habitual, lived body and the broken, alienated body allows the assumption that biomedical science only plays a part in the constitution of our embodied being-in-the-world after the body has broken. Hence it is presupposed that the ethical dimension of biomedicine is restricted to its practice of altering bodies in the *repair* of our 'social expressiveness'. However, biomedicine is not just a practice which works directly on bodies. It is also one of the discourses which make up the world in which we ordinarily dwell. Biomedicine is a field of knowledge and, as Michel Foucault suggests, there is no field of 'knowledge that does not presuppose and constitute at the same time power relations' (Foucault 1979: 27). Further, biomedicine is not just one among many fields of knowledge which regulate bodies in the name of the so-called common good: it holds a privileged place in disseminating knowledge about what a body is, how it functions and the nature of its capabilities. And, in this, biomedical knowledge does its own social dichotomising in delineating the normal body from the abnormal. So it is possible not only that biomedical science is involved in the restoration and expansion of bodies upon which it practices, but that as a field of knowledge it may play a part in the constitution of those bodies prior to any alienation. Consequently, the ethics of biomedicine may be located not just within its practice of restoring or enhancing our embodied being-in-the-world, but also in that which regulates our wellbeing and hence our embodied sexed identity.

In chapter 2 I outlined, with reference to the work of Foucault, how both biomedical knowledge and biomedical ethics may be implicated in the constitution and normalisation of sexed bodies. I argued that the nexus between concerns about health and concerns about a productive labour force effectively divides the healthy from the unhealthy in terms of what kinds of bodies are deemed useful to our social and economic system. Through surveillance, discipline and confession, with reference to knowledge about the health of bodies, bodies are transformed as they are distributed around a norm. I suggested that the operation of this productive power is such that the maternal body is a privileged target of medical and moral attention because it is the site of the reproduction of the social body. Through this attention, caught as it is within the political investment of 'healthy' productive bodies suitable for exchange in a labour market, the maternal body is constituted as other to the 'normal' body.

The analyses of sexual difference offered in the intervening chapters, and some feminist work on biomedical science not yet mentioned, allow me to be more specific about this operation of productive power and the role biomedical science plays within it regarding the constitution of sexed bodies. The first question to be addressed is what is the biomedical model of the human body? Donna Haraway (1985) has suggested that developments in communication and information technologies have changed our conception of the human body from a passive mechanism which houses a consciousness to an information system (or cyborg). And biomedical science has been instrumental in developing, legitimating and refining this communication model of both the human body and its internal (genetic, neurological, endocrine and immune) systems (Haraway 1989).

Of particular interest here is how sexed bodies are represented within this knowledge about the body and to what effect. Any brief excursion into an anatomy and physiology text will reveal that the 'normal' body of biomedical science is male. But beyond this by now well-rehearsed point, the 'normal' biomedical body is represented as an ordered system of coded units (genes, cells, organs) and of information flows across and between the boundaries of these units. This body is said to be 'a hierarchically organised bureaucratic system of control', as Emily Martin puts it (Martin 1990: 74), where all communication (genetic, neurological, hormonal) is said to have a purpose. As a consequence of this model, illness, internal corporeal change and differences between bodies are understood in terms of coding errors and information malfunction (Haraway 1989: 15) or in terms of loss of control and failure of purpose (Martin 1990: 75). This microscopic constitution of deviance is particularly evident in the biomedical representation of bodily functions peculiar to women where menstruation, for example, is represented as a failure to achieve the said goal of a woman's endocrine system: to provide the conditions necessary to support a fertilised human ovum (Martin 1990: 74–5). Given that corporeal change and difference is understood in terms of error, malfunction, deviation, then, as Haraway suggests, sexual and racial discrimination is now more likely to occur, not with direct recourse to claims about women's reproductive role or about the primitivism of other cultures, but by evoking parameters for correct coding and proper and efficient information flow within and between bodies (Haraway 1985: 81). Biomedical knowledge about bodies affects and feeds into an almost invisible programme of injustice without the presence of an intentional agent of power.

What is not immediately clear from these accounts is *why* illness, corporeal change and sexual difference emerge from this information model of the body as deviance; why and how this translates into discrimination at a macro level; and whether this mode of representation has material effects aside from those due to what appears to be supplementary injustice. Regarding the question of constitutive effects, Martin implies that if there are any, this is at the level of how women represent rather than experience their bodies. And she suggests that because working-class women, in their self-representations, resist the medical model of menstruation, for example, then it is possible for middle-class women (who do accept the model) to learn to 'escape the prevailing scientific view' (Martin 1990: 79). Haraway is less inclined to separate lived experience from its representation, and without assuming an escape route finds positive possibilities in the constitutive effects of biomedical knowledge of the body. She suggests that biomedical knowledge 'maps our bodily reality' by 'fracturing identities' and encouraging new affinities (Haraway 1985: 66, 73). As to the mechanism of this opening of new possibilities, Haraway later suggests that this fracturing of identity is an effect of the way in which biomedical discourses 'have destabilized the symbolic privilege of the hierarchical localized, organic body' at the same time as 'the question of "differences" has destabilized humanist discourses of liberation based on a politics of identity and substantive unity' (Haraway 1989: 14).

Without denying that the biomedical model of the body is saturated with coding, information and teleological metaphors and that it has both discriminatory and uncertain effects, I want to propose that the itinerary of its constitutive effects is of a different order to, although not altogether inconsistent with, that proposed by Haraway and Martin. As I have argued throughout, the 'question of differences' (raised against the humanist model of identity) does *not* imply that identity is lived as fractured. Rather these interventions on difference suggest that, insofar as privileged identity is said to be and is lived as a substantial unity, others are constituted as fractured and deviant and are marginalised and silenced as a result.

It is within the terms of this assumption of atomised, unified identity, rather than via a destabilisation of it, that the biomedical model of the body has its constitutive effects. That is, it is precisely because the biomedical model of the 'normal' body is given as a 'hierarchical, localised, organic body' that difference is viewed and lived as deviation, error and malfunction and discrimination can operate on these grounds. My suggestion is that what is problematic about the biomedical model of the body is its implicit understanding of identity and difference. It is by virtue of this understanding that biomedical science is deeply implicated in the intellectual rigidity discussed above. By treating bodies as self-contained and as either absolutely different or identical, it effaces as it contributes to the ambiguity inherent in embodied existence. And my suggestion is that insofar as this rigidity, and any attendant injustice, is apparent in the practice of biomedicine, the preconditions are already there in theory.

At a general level, the biomedical model of the body shares assumptions with the liberal and contract models of the self already discussed: both effectively separate the self from the body and reduce the body to a passive material object isolated from other bodies in space. I have already noted how assuming the body is a mere appendage to the self encourages the objectification of the self apparent in medical practice. But the atomism of the biomedical model of the body has further consequences. By assuming that the body is given in space separate from its world, biomedical science can claim to know its identity with mathematical precision. Accompanying the claim to know a body's identity is the claim to know, at least potentially, in what ways and for what reasons bodies differ. In this there is no appreciation of the social interrelations between bodies, the constitutive effects of those relations and the ambiguities inherent in them. And, more important, there is no acknowledgement of the constitutive relation between the knower and the object of knowledge.

Biomedical science owes its status as a science to an absolute distinction between the subject of knowledge and its object. However, as I have suggested, following Merleau-Ponty, any lived distance between the subject and the body of another is constituted with an attendant imposition of one's corporeal schema onto the other. Biomedical science, in claiming objectivity and by assuming the singularity of the body it attempts to know, denies such constitutive effects. After differentiating itself from the dwelling of the other, it deals with the strangeness of the other's body either by reducing it to something already familiar ('the task of knowledge consists in making the other become the Same' (Levinas 1985: 91)) or by reducing it to a negative image of the same.

Through its atomised model of identity, biomedical science builds a single model of a singular body with reference to familiar assumptions about what is proper to bodies. And, as constituting any body as proper and self-contained involves the negation of ambiguity, of thoughtless gifts, differences can only be understood in terms of deviation from this norm. Given this model of identity and difference, it should not be surprising if the general tendency to identify women's bodies against a male norm, already discussed in detail, gets reproduced within this scientific knowledge of the body. Nor should it be surprising if this constitution of deviance is reproduced in practice: in the treatment of pregnant bodies as disabled, for example.

Of course, scientific knowledge of the human body is less concerned with the macroscopic body than it is with the microscopic workings of its inner life.[7] It is in the representation of the body's inner life that we find those information and teleological metaphors of concern to Haraway and Martin. These information models of the body's genetic, endocrine, neurological and immune systems, far from destabilising the atomistic model of identity just discussed, actually reinforce its terms. The idea of a singular body, given in isolation from other bodies and from the knowledge which describes it, is reproduced in the representation of genes, cells, neurones, organs and their relations. While these relations are given in terms of transmission of information, messages are said to move between singular unities and across discrete borders. There is no challenge to unified identity here. In fact, the whole enterprise of biomedical science consists in a search for the origin of identity, the microscopic origin of the appearances of, and differences between, macroscopic human bodies.

I have argued elsewhere, in the context of an account of genetic theory (Diprose 1991), that this search for origins is mediated by and reinforces the same atomistic individualism and social dichotimising that governs social relations between bodies at the macro level. In that analysis I suggested that biomedical science posits the genetic code as the origin of bodily identity. That biomedical science is inhabited by a familiar and problematic model of unified identity is indicated by its accounts of the pathway between the genetic code and its expression: this is said to involve the exact translation of an original, discrete code, through the production of its mirror-image, into its (protein) double, and so on. It is because this information model insists on a locatable origin of identity and assumes exact translation between the original code and its expression that difference, illness and corporeal change are understood in terms of disruption, breakdown or mutation in the process of transmission. This representation of identity and difference mirrors and reinforces that assumed in practical genetics where the ethicist is justifiably concerned about the potential effacement of differences in the name of what is assumed to be proper to bodies (Diprose 1991: 70). So, if sexual discrimination can occur, say by evoking parameters of genetic coding errors, this is predicated upon a problematic model of identity and difference in genetic theory, one which seeks to secure, rather than challenge, a humanist model of unified self-identity.

The biomedical scientist does acknowledge that the information flow which is said to regulate the body is governed by uncertainties. However, this uncertainty is put down to the limits of knowledge. What I would suggest is that the uncertainty and ambiguity apparent in the scientific encounter with the human body are the effect of its practice of interpretation rather than a mark of the limits of its knowledge. Cathryn Vasseleu (1991) suggests that biomedical science, in addressing the limits of its knowledge, appeals to the idea that its models of the body are only metaphors. The same distinction, between the body and the metaphors which describe it, is borrowed by those critics of science who, in pointing to the discriminatory effects of these metaphors, seek to change or escape them. However, Vasseleu argues, what is overlooked in appealing to a distinction between the body and its representations is that the distinction legitimates the scientific enterprise ('[s]cience preserves "life itself" as a domain in need of constant reformulation by *it*') as it obscures the material effects of that enterprise ('[t]o read the biological body as simply the essential body is to ignore the essentialising function of . . . the body-writing practices of biomedical science') (Vasseleu 1991: 64).

These models and metaphors inform not just the way we represent our bodies to ourselves, but how we live our bodies. And these effects are not simply added to a natural body which could slip out from under its scientific representation. As I have argued in previous chapters, and here with reference to Merleau-Ponty, the lived body which we are is constituted and maintained through an ongoing differentiation from others, through the constitution of a lived distance or difference from other bodies. Insofar as biomedical knowledge of the body enjoys the status of a science, it provides the dominant terms of this differentiation process through its formative mapping of the borders between the genetic code and its expression, between genes, between other microscopic units of identity, between the subject of knowledge and its body-object and between macroscopic human bodies.

On the one hand, biomedical science makes micro and macroscopic differences real by the use of metaphors which produce singular identities and organise differences through relations on the basis of sameness. In this it echoes the wider habit of marking male bodies off as proper unities by rendering women improper and disabled. On the other hand, this process of production is inhabited by the uncertainty, ambiguity or residue apparent in the creation of any ethos. This residue is not the reality of a woman's body somehow hidden by its scientific representation. Rather it is, in Merleau-Ponty's terms, the interworld that is the lived body: that effect of the way in which the partitioning off of a singular body, by scientific or other means of objectifying a body, is always marked by that body's reference to another body.[8] Or, to put this otherwise, a woman's body does not coincide with its biomedical representation because, just as scientific representation of bodies constitutes borders within and between bodies and between its object and itself, it also guarantees the interruption of self-identity, the dispersal of every assemblage of self (Diprose 1991: 75). This not only suggests that biomedical science will have difficulty catching up with the bodies it constitutes, it also suggests other possibilities for the actualisation of women's biologies beyond failure, breakdown and error. At the same time it is important to remember that, as these other possibilities are opened within knowledges which privilege a unified male body, they are more likely to be lived by women, under present conditions, as disabling.

If ethics is understood as the problematic of the constitution of sexed, embodied ethos, then it would seem prudent to include within this problematic the scientific representations of sexed bodies. If biomedical science, as a field of knowledge, is implicated in the production of sexed identity and difference, the ethics of biomedical science does not begin with its practice of fixing 'broken' bodies. Nor does it begin with the objectification of women's bodies in the consulting room, research clinic or hospital ward. Rather the ethics of biomedical science begins with its knowledge of bodies through which it is not only complicit in the regulation of the wellbeing of women's bodies but may also provide the conditions for the possibility of their brokenness.

Chapter 7

Conclusions

Ethics, understood as the interrogative practice of that which constitutes our sexed and embodied place in the world, is not positive in the sense of providing rules for action or a blueprint for change. The reader who has been searching for a guide to the good life will not find one here. And this is because, as I have attempted to demonstrate, a positive ethics, which assumes or aims for a common good, can do more harm in a day than any transgressor of a moral code could do in a lifetime. In the midst of complex and fragmentary patterns of modern life, we may desire the stability of a place of our own. And faced with almost invisible forms of injustice we may seek comfort and security in familiar rules and entrenched modes of regulation. Yet such a place is what dreams are made of and such security is assumed at a cost. The regimes of social regulation, which dictate the right way to live, implicitly or explicitly seek to preserve the integrity of every body such that we are compatible with the social body. Not only do these thereby dictate which embodied existences can be transformed by whom and to what end, but, as it is here that comparisons are made and values born, not all bodies are counted as socially viable. In short, the privilege of a stable place within that social and political place we call the 'common good' is secured at the cost of denigrating and excluding others.

While not positive, the ethics I have proposed is not negative. It is not a nihilistic call for the suspension of all social regulation, nor does it imply withdrawal into inaction. As I have argued, insofar as we are the embodied products of regimes which regulate sexual difference, these regimes support our existence. Thus, there is no back door to freedom via anarchy or solitude.

While liberty may not be possible, change is. As I have also argued, no existence is determined in-itself, definitively. Those comparative modes of evaluation which constitute and regulate life and position woman as 'other' to man also divide existence, ensuring that no sexed identity can be contained or confined to one place. It is the task of an ethics of sexual difference to build other possibilities from what already is, from the ambiguities and contradictions inherent in the containment of women as men's 'others'.

In this task we all become the subject and object of ethics: the judges of those normative discourses which distribute injustice but also the vehicles through which they do their work. And in interrogating the regimes which regulate sexual difference, the object of ethics is not simply those value-laden discourses with an explicit moral code, nor just those philosophical discourses which describe, justify and underpin hierarchical social and political structures. Additionally, and perhaps most urgently, ethics must include those scientific representations of embodied identity and difference which, in claiming objectivity, appear to be devoid of a moral and political agenda and free of material effects. Without ignoring those familiar instances of injustice against women, it is also necessary to pay attention to the network of discursive mechanisms which, by constituting our relative places in the world, ground injustice and make it possible.

Notes

CHAPTER 1 FEMINISM AND THE ETHICS OF REPRODUCTION

1 In her submission Dr Diana Kirby (Melbourne Feminist Legal Therapy Group) argues against the surrogacy contract on the grounds that a surrogate mother cannot predict the strength of her emotional attachment to the child at the time of birth. This inability to predict one's future wellbeing is a common feminist argument against the surrogacy contract. See, for example, Dodds and Jones (1989).

2 Susan Sherwin (1989), for example, formulates this kind of 'contextual' framework for evaluating abortion, reproductive technology and surrogacy.

3 For critical discussions of this debate see, for example, various essays in Eva Feder Kittay and Diana T. Meyers (eds), *Women and Moral Theory* (1987); Eve Browning Cole and Susan Coultrap-McQuin (eds), *Explorations in Feminist Ethics* (1992); Seyla Benhabib, *Situating the Self* (1992: chs. 5 and 6); Elizabeth Porter, *Women and Moral Identity* (1991: ch. 6) and Lawrence Blum (1988).

4 For an early critique of Gilligan on this point see Jean Grimshaw, *Feminist Philosophers* (1986: ch. 7) and for a recent discussion see Patricia Ward Scaltsas (1992).

5 Both Porter and Benhabib do develop this aspect of their theses but in ways I find problematic. Benhabib, while acknowledging the interrelation of self and other, bases the possibility of a 'moral conversation' on the separation of the other's identity from that of the self (Benhabib 1992: 10). Porter posits a dialectical synthesis of moral identities which she claims will retain differences (Porter 1991: 49). As I will argue in chapters 3 and 4, neither kind of position captures the complexities of identity and difference.

CHAPTER 2 ETHICS, EMBODIMENT AND SEXUAL DIFFERENCE

1 See, for example, Charles Scott's etymology of 'ethos' in 'Heidegger and the Question of Ethics' (Scott 1988: 23–8). Scott has gone on to explore this alternative understanding of ethics through analyses of the ethics of Nietzsche, Foucault and Heidegger (Scott 1990).

2 For example, 'ethos' is defined by Aristotle in his *Nicomachean Ethics*, bk 2, ch. 1 as character established through habitual action.

3 The meaning of 'being-in' is Martin Heidegger's question in *Being and Time* (1962). See in particular pp. 78–86.

4 See, for example, the work of Michel Foucault; Luce Irigaray, particularly *Ethique de la Différence Sexuelle* (1984; partially reprinted in Irigaray 1991);

Moira Gatens (1988 and 1989); Rosi Braidotti (1986); and Emmanuel Levinas particularly *Totality and Infinity* (1969).

5 Power, according to Foucault, is not so much a relation which links two parties to the absolute detriment of the freedom of one, as a more invisible mode of 'government' where 'to govern' means 'to structure the possible field of action of others'. Therefore, power is exercised over those who are 'free' to act within a range of possibilities. The exercise of power determines the possibilities of action. See, for example, Foucault (1982). For a comprehensive discussion of Foucault's model of power see Paul Patton (1989).

6 Foucault traces the rise of this political rationality, which he calls the 'pastoral concept of Government', in 'Omnes et singulatim: towards a criticism of political reason' (Foucault 1981).

7 'Discipline "makes" individuals; it is a specific technique of power that regards individuals both as objects and as instruments of its exercise' (Foucault 1979: 170). Foucault uses the 'panopticon', Jeremy Bentham's proposed system of incarceration, as a model of the productive operation of power. Within this economy of power, the individual 'becomes the principle of his own subjection' (ibid.: 203).

8 For various accounts of the social production of women's bodies which use Foucault's model of productive power see Diamond and Quinby (eds) (1988).

9 Bordo's analyses of the cultural production of women's modes of embodiment (see also Bordo: 1990), while explicitly indebted to Foucault, necessarily depart from him on the possibility of an aesthetics of self for women under current cultural conditions.

10 Foucault says little more than this about the exclusion of women, implying that the exclusions are incidental to Ancient Greek ethics. However, as I have argued elsewhere (Diprose 1989), the value attributed to the Ancient Greek ethical subject is generated by the exclusion of women and others.

11 For a description of this phenomenon in the Australian workplace, see Game and Pringle (1983).

12 For an alternative and illuminating account of how Irigaray's work can be read as a critique of the social contract and its systematic exclusion of women see Whitford (1991: ch. 8).

13 For discussions and defences of Irigaray's work against the charge of essentialism see Grosz (1989), Kirby (1991) and Whitford (1991).

CHAPTER 3 HEGEL'S RESTRICTED ECONOMY OF DIFFERENCE

1 Hegel's explicit theory of the sign can be found in the *Philosophy of Mind* sections 451–64. However, the idea contained there, that the meaning of a sign is derived from its relation to what it is not, pervades all of his work.

2 Hegel posits this idea most clearly in his analysis of 'sense certainty' in the *Phenomenology of Spirit*, sections 90–110.

3 Jacques Derrida (1982), for example, in his reading of Hegel's theory of the sign, has demonstrated how representation, for Hegel, arises from, as it produces, a distinction between thought and matter (between conscious ideas and their material signifiers). The difference between thought and that which is transformed to represent it is both a necessary condition of representation and produced within representation. For Hegel, the difference is supposedly dissolved in the sign. Yet Derrida finds, within Hegel's own account, that neither thought nor its signifier comes before the difference which brings them into existence, and this difference is produced rather than effaced in

the sign: 'the opposition of soul and body, and analogically the opposition of the intelligible and the sensory, condition the difference between the signifying intention (*bedeuten*), which is an animating activity, and the inert body of the signifier' (Derrida 1982: 82). 'This heterogeneity amounts to the irreducibility of the soul and the body, of the intelligible and the sensory, of the concept and the signified ideality on the one hand, and of the signifying body on the other, that is, in different senses, the irreducibility of two *representations* (*Vorstellungen*)' (ibid.: 84). Irene Harvey provides an account of Derrida's critique of Hegel on this point (Harvey 1986: 20).

4 I use the pronoun 'he' in the discussion of habits because, as I will go on to argue, the proper body in Hegel's ethical life is male.

5 I note Butler's reading of Hegel on desire because, unlike other commentators, she carefully traces what happens to the self's embodiment in the social transformation of difference into identity. This is important for establishing the material effects of the representation of sexual difference in ethical life. Butler does not tackle this latter issue but her reading is suggestive and important.

6 Hegel concedes, for example, at the end of the dialectic of self-consciousness, that the work of the 'Bondsman' in the presence of the 'Lord' does not achieve the unity of identity and difference. We are left with 'unhappy consciousness' where two contradictory moments of selfhood (the unchangeable, self-identical 'I' and consciousness of the contingency of life – due to one's embodiment and one's relation to the other) are combined into one. This 'unhappy consciousness' is an awareness of the self as a duality. Some commentators of Hegel, Hyppolite for example, view 'unhappy consciousness' as the essence of subjectivity (Hyppolite 1974: 203). Hegel, however, persists in assuming that this alienation of the self in the other can be overcome.

7 For a discussion of how Hegel sets up rational consciousness in opposition to the 'feminine' and makes the political domain dependent on women's exclusion see Genevieve Lloyd (1984: 70–85).

8 For a discussion of how the relation between war and masculinity, as suggested by Hegel, still affects women in contemporary social relations see Lloyd (1986).

9 I am quoting from the *Philosophy of Right* here rather than the *Phenomenology of Spirit*. The treatment of sexual difference is slightly different in the two texts in that the former deals with the family in modernity, the latter in antiquity. However, both are based on the same anthropology and ontology which is what matters in my analysis. My use of both texts relies on these similarities without collapsing the differences. For a detailed discussion of Hegel's treatment of ethics in the *Philosophy of Right* see, for example, Wood (1990).

10 The connection between these two passages, although not my conclusions about the connection, is suggested by Hodge (1987).

11 Although de Beauvoir is also aware that part of the cause of women's alienation is the way in which women's embodiment is socially represented. For discussions of the complexities and ambiguities of de Beauvoir's analysis of sexual difference see Gatens (1991a: 48–59), Lloyd (1984: 93–102) and Mackenzie (1986).

12 Hodge's analysis of the tensions in Hegel's philosophy, generated by the exclusion of women, is thorough and astute. However, as I will go on to argue, I doubt whether her conclusion will remove the tensions she describes.

CHAPTER 4 SEXUAL DIFFERENCE BEYOND DUALITY

1 Derrida, 'Force of Law: The "Mystical Foundations of Authority" ', *Cardozo Law Review*, vol. 11, nos. 5–6, 1990. Quoted in Cornell (1991: 112).

2 At least this is where Derrida (1983), analysing the distinction Heidegger makes between the sexual 'neutrality' of *Dasein's* dispersed structure and its parcelling out into the two sexes, claims Heidegger locates sexual difference as duality.

3 Insofar as Heidegger insists on some kind of authentic understanding relation to Being, uncontaminated by the 'they', he can be and has been charged with promoting metaphysical egoism. Bernasconi (1988) discusses the consequences of Heidegger's insistence on maintaining this priority of the ontological dispersed structure of *Dasein* over its ontic manifestations ('differences' derived from its primordial dispersed structure) and ontic modes of inquiry (such as an ethics based on an ontology of presence). Bernasconi also stresses the difficulty Heidegger has in maintaining the purity of the distinctions between ontological and ontical, authenticity and inauthenticity.

4 The immediate problem with equating authenticity with the feminine is that it ignores Heidegger's claim that the primordial dispersed structure of *Dasein* is sexless: 'The peculiar *neutrality* of the term "Dasein" is essential, because the interpretation of this being must be considered prior to every factual concretion. This neutrality also indicates that Dasein is neither of the two sexes' (Heidegger 1984: 136). Of course we do not have to accept this claim of sexual neutrality. Derrida argues that, as this primordial structure is one of ontological difference rather than unity and homogeneity, then it must be marked by sexual difference (beyond and necessary to sexual duality):

> If *Dasein* as such belongs to neither of the two sexes, that doesn't mean that its being is deprived of sex. On the contrary, here one must think of a pre-differential, rather a pre-dual, sexuality – which doesn't mean unitary, homogeneous, or undifferentiated.
>
> (Derrida 1983: 72)

Derrida demonstrates that the 'neutrality' of *Dasein* is not a (Hegelian) neutrality or unity in opposition to binary sexual difference. Rather, this supposed neutrality opens a way to think a sexual difference which exceeds the negativity of sexual duality. However, this is not the same as saying *Dasein's* primordial structure is feminine as opposed to (masculine) self-presence.

5 'The useful part of deconstruction is in the suggestion that the subject is always centered. Deconstruction persistently notices that this centering is an effect-structure with indeterminable boundaries that can only be deciphered as determining', Gayatri Chakravorty Spivak (1989: 214). Also, Spivak argues elsewhere that Derrida's 'point is precisely that man can problematize but not fully disown his status as subject' (Spivak 1982: 178).

CHAPTER 5 NIETZSCHE ON SEXED EMBODIMENT

1 Nietzsche makes a similar comment about the derivative nature of pleasure and pain in *Beyond Good and Evil* (Nietzsche 1972: 135–6).

2 For a discussion of the problem of the actor in Nietzsche's philosophy see Paul Patton (1991b).

3 The idea that the bodily self is reproduced differently as it is temporalised

through the production of a distance within the self would seem to be at odds with Nietzsche's doctrine of eternal recurrence. Problems arise if we accept eternal recurrence as either a cosmological hypothesis, where the world repeats itself infinitely (Nietzsche 1967: 521), or a psychological doctrine, where self-affirmation involves the desire for the self to recur eternally the same (Nietzsche 1978: 322). However, as David Wood (1988) has demonstrated, interpreting the doctrine of eternal recurrence exclusively in either of these ways is ultimately untenable. Besides the cosmological and psychological doctrines of eternal recurrence, Wood discusses a third possible interpretation, the 'ontological', which I have found useful. For my own interpretation of Nietzsche's presentation of the doctrine in 'The Vision and the Riddle' (Nietzsche 1978: 155–60) see Diprose (1993: 7–8).

4 For discussions of Nietzsche's use of the metaphor of pregnancy see Alison Ainley (1988) and Paul Patton (1991a: 49–52).

5 Nietzsche makes a further connection between interpretation and will to power as a form giving force in *On the Genealogy of Morals* (Nietzsche 1969: 79).

6 Nietzsche makes similar observations on the disjunction between self-interpretation and interpretation by another in *Beyond Good and Evil* (Nietzsche 1972: 97, 142).

7 For Nietzsche's understanding of the different ways a man can possess a woman and what these say about the man's self-image see Nietzsche (1972: 98–9).

8 I discuss Nietzsche's opposition to feminism of equality in more detail elsewhere (Diprose 1989).

9 Nietzsche's claim that women put on something when they take off everything has often been interpreted as faking orgasm – woman's constitution of her own self-presence when appearing to guarantee man's. Or, as Gayatri Spivak suggests: 'Women, "acting out" their pleasure in the orgasmic moment, can cite themselves in their very self-presence' (1984: 22). I take issue with Spivak only in her claim that it is self-presence (rather than undecidable difference) which is being cited in woman's dissimulation.

CHAPTER 6 BIOMEDICAL ETHICS AND LIVED, SEXED BODIES

1 I am grateful to Deborah Keighley-James for bringing this paper to my attention.

2 An objection may be raised at this point to the effect that Merleau-Ponty's model of embodied being-in-the-world does not include those who are relatively inactive: a quadriplegic or an intellectual for example. His answer would be that insofar as both are absorbed in a situation and exist in a familiar dwelling with their bodies in their world, then both are body-subjects with intentionality and future possibilities (Merleau-Ponty 1962: 164–5, 435).

3 Merleau-Ponty discusses this self-conscious mode of being in the world using the case of Schneider, who can perform habitual actions but not new purposeful actions (1962: 103–20). When asked to trace a circle in the air, for example, Schneider needs to deliberate, take his body and its parts as an object and gradually map out the action through a series of disjointed movements until he reaches the one required. For a detailed discussion of Merleau-Ponty's use of this case, see Hammond *et al.* (1991).

4 For a discussion of the ethics of abortion which partly follows the existential phenomenological tradition see Mackenzie (1992).

5 Dodds and Jones are not alone in presenting this kind of argument against surrogacy but their discussion stands out as being particularly thorough and clear. While I go on to refute their claims, so do they. In an equally thorough analysis, Dodds and Jones (1992) have since rethought their position, arguing that it is not so much surrogacy which is the problem but the surrogacy contract and the contract model of the individual.

6 Jennifer Biddle, with partial reference to Merleau-Ponty, provides an illuminating analysis of this dis-ease in the context of the anthropologist's experience of a different culture. She argues that high rates of illness and accidents among anthropologists in the field can be understood as the effect of the necessary reconstitution of one's embodied existence when dwelling with others whose bodies are initially foreign to the self. My use of Merleau-Ponty for an analysis of sexual difference has been enriched by numerous discussions with Jennifer for which I am grateful.

Iris Young (1990b: 122–55), with reference to Kristeva's notion of abjection, accounts for racism, homophobia and sexism in terms of an embodied aversion to bodies whose difference threatens the integrity of the self. As she suggests, the tendency to alleviate this anxiety by denigrating the other is masked rather than redressed by a conscious discourse of equality.

7 Although, as I will go on to suggest, there is an intimate connection between social assumptions about the relations and differences between macroscopic human bodies and the scientific models of the body's inner life. For a more detailed discussion of this relation, and its constitutive effects, with particular reference to the relation between discrimination against homosexuals and the scientific model of the aetiology of AIDS, see Diprose and Vasseleu (1991).

8 The idea that knowledge of the body is implicated in the building of partitions between bodies puts a question mark over Merleau-Ponty's suggestion that intellectual knowing of the other's body is secondary to 'syncretic sociability' and is a deficient mode of being-with-others. While I do not have space to elaborate here, my analysis would suggest that if sociability involves a lived distance between self and other and if this is based on the objectification of bodies in which science is implicated, then scientific knowledge of bodies sets the terms for what occurs in syncretic sociability, at least in part.

References

Ainley, Alison (1988) ' "Ideal selfishness": Nietzsche's metaphor of maternity', in David Farrell Krell and David Wood (eds) *Exceedingly Nietzsche: Aspects of Contemporary Nietzsche-Interpretation*, London and New York: Routledge.

Aristotle (1953) *The Nicomachean Ethics*, published as *The Ethics of Aristotle*, trans. J.A.K. Thomson, Harmondsworth, Middlesex: Penguin.

de Beauvoir, Simone (1972) *The Second Sex*, trans. H.M. Parshley, Harmondsworth, Middlesex: Penguin.

Benhabib, Seyla (1992) *Situating the Self: Gender, Community and Postmodernism in Contemporary Ethics*, Cambridge: Polity.

Bernasconi, Robert (1988) ' "The double concept of philosophy" and the place of ethics in *Being and Time*', *Research in Phenomenology* 18: 41–57.

Biddle, Jennifer (work in progress) 'The anthropologist's body', University of Sydney.

Blum, Lawrence (1988) 'Gilligan and Kohlberg: implications for moral theory', *Ethics* 98, 3: 488–9.

Bordo, Susan R. (1989) 'The body and the reproduction of femininity: a feminist appropriation of Foucault', in Alison M. Jagger and Susan R. Bordo (eds) *Gender/Body/Knowledge: Feminist Reconstructions of Being and Knowing*, New Brunswick: Rutgers University Press.

—— (1990) 'Reading the slender body', in Mary Jacobus, Evelyn Fox Keller and Sally Shuttleworth (eds) *Body/Politics: Women and the Discourses of Science*, New York and London: Routledge.

Braidotti, Rosi (1986) 'Ethics revisited: women and/in philosophy', in Carole Pateman and Elizabeth Gross (eds) *Feminist Challenges: Social and Political Theory*, Sydney: Allen & Unwin.

—— (1989) 'The politics of ontological difference', in Teresa Brennan (ed.) *Between Feminism and Psychoanalysis*, London and New York: Routledge.

—— (1991) *Patterns of Dissonance: A Study of Women in Contemporary Philosophy*, trans. Elizabeth Guild, Cambridge: Polity.

Brennan, Teresa (ed.) (1989) *Between Feminism and Psychoanalysis*, London and New York: Routledge.

Butler, Judith P. (1987) *Subjects of Desire: Hegelian Reflections in Twentieth-Century France*, New York: Columbia University Press.

—— (1990) *Gender Trouble: Feminism and the Subversion of Identity*, New York and London: Routledge.

Callahan, Daniel (1984) 'Autonomy: A Moral Good, not a Moral Obsession', *The Hastings Centre Report*, October: 40–2.

Chodorow, Nancy (1978) *The Reproduction of Mothering: Psychoanalysis and the Sociology of Gender*, Berkeley: University of California Press.

Cole, Eve Browning and Coultrap-McQuinn, Susan (eds) (1992) *Explorations in Feminist Ethics: Theory and Practice*, Bloomington: Indiana University Press.

Cornell, Drucilla (1991) *Beyond Accommodation: Ethical Feminism, Deconstruction and the Law*, New York and London: Routledge.

Derrida, Jacques (1974) *Of Grammatology*, trans. Gayatri Chakravorti Spivak, Baltimore: The Johns Hopkins University Press.

—— (1978) *Writing and Difference*, trans. Alan Bass, Chicago: University of Chicago Press.

—— (1979) *Spurs: Nietzsche's Styles*, trans. Alan Bass, Chicago: University of Chicago Press.

—— (1981a) *Dissemination*, trans. Barbara Johnson, Chicago: University of Chicago Press.

—— (1981b) *Positions*, trans. Alan Bass, Chicago: University of Chicago Press.

—— (1982) *Margins of Philosophy*, trans. Alan Bass, Chicago: University of Chicago Press.

—— (1983) 'Geschlecht: sexual difference, ontological difference', *Research in Phenomenology* 13: 65–83.

—— (1985) 'Choreographies' an interview with Christie McDonald, in Christie McDonald (ed.) *The Ear of the Other*, trans. Peggy Kamuf, Lincoln: University of Nebraska Press.

—— (1986) *Glas*, trans. John P. Leavey and Richard Rand, Lincoln: University of Nebraska Press.

—— (1987) 'Women in the beehive: a seminar with Jacques Derrida', in Alice Jardine and Paul Smith (eds) *Men in Feminism*, New York: Methuen.

—— (1992) 'Given time: the time of the king', *Critical Inquiry* 18, Winter: 161–87.

Diamond, Irene and Quinby, Lee (eds) (1988) *Foucault and Feminism: Reflections and Resistance*, Boston: Northeastern University Press.

Diprose, Rosalyn (1987) 'The use of pleasure in the constitution of the body', *Australian Feminist Studies* 5: 95–103.

—— (1989) 'Nietzsche, ethics and sexual difference', *Radical Philosophy* 52, Summer: 27–33.

—— (1991) 'A "Genethics" that makes sense', in Rosalyn Diprose and Robyn Ferrell (eds) *Cartographies: Poststructuralism and the Mapping of Bodies and Spaces*, Sydney: Allen & Unwin.

—— (1993) 'Nietzsche and the pathos of distance', in Paul Patton (ed.) *Nietzsche, Feminism and Political Theory*, London: Routledge.

Diprose, Rosalyn and Ferrell, Robyn (eds) (1991) *Cartographies: Poststructuralism and the Mapping of Bodies and Spaces*, Sydney: Allen & Unwin.

Diprose, Rosalyn and Vasseleu, Cathryn (1991) 'Animation-AIDS in science/fiction', in Alan Cholodenko (ed.) *The Illusion of Life: Essays on Animation*, Sydney: Power Institute.

Dodds, Susan and Jones, Karen (1989) 'Surrogacy and autonomy', *Bioethics* 3, 1: 1–17.

—— (1992) 'Surrogacy and the body as property', in Stephen Darling (ed.) *Cross-Currents: Philosophy in the Nineties*, Adelaide: Flinders Press.

Ferrell, Robyn (1993) 'Why bother? Defending Derrida and the significance of writing', *Australasian Journal of Philosophy* 71, 2, June: 121–31.

Foucault, Michel (1978) *The History of Sexuality Volume One: an Introduction*, trans. Robert Hurley, New York: Random House.

—— (1979) *Discipline and Punish: the Birth of the Prison*, trans. Alan Sheridan, Harmondsworth, Middlesex: Penguin.

—— (1980) *Power/Knowledge: Selected Interviews and Other Writings 1972–1977* ed. Colin Gordon, Brighton, Sussex: Harvester Press.

—— (1981) 'Omnes et singulatim: towards a criticism of political reason' in M. McMurrin (ed.) *The Tanner Lectures on Human Values Volume II*, Salt Lake City: University of Utah Press.

—— (1982) 'The subject of power', afterword to Hubert L. Dreyfus and Paul Rabinow *Michel Foucault: Beyond Structuralism and Hermeneutics*, Brighton, Sussex: Harvester Press.

—— (1984) *The Foucault Reader* ed. Paul Rabinow, Harmondsworth, Middlesex: Penguin.

—— (1987) *The Use of Pleasure: The History of Sexuality Volume Two*, trans. Robert Hurley, Harmondsworth, Middlesex: Penguin.

Game, Ann and Pringle, Rosemary (1983) *Gender at Work*, Sydney: Allen & Unwin.

Gatens, Moira (1988) 'Towards a feminist philosophy of the body', in Barbara Caine, E.A. Grosz and Marie de Lepervanche (eds) *Crossing Boundaries: Feminisms and the Critique of Knowledges*, Sydney: Allen & Unwin.

—— (1989) 'Women and her double(s): sex, gender and ethics', *Australian Feminist Studies* 10: 33–47.

—— (1991a) *Feminism and Philosophy: Perspectives on Difference and Equality*, Cambridge: Polity.

—— (1991b) 'Corporeal representation in/and the body politic', in Rosalyn Diprose and Robyn Ferrell (eds) *Cartographies: Poststructuralism and the Mapping of Bodies and Spaces*, Sydney: Allen & Unwin.

Gilligan, Carol (1982) *In a Different Voice: Psychological Theory and Women's Development*, Cambridge, Mass.: Harvard University Press.

Graybeal, Jean (1990) *Language and 'The Feminine' in Nietzsche and Heidegger*, Bloomington: Indiana University Press.

Grimshaw, Jean (1986) *Feminist Philosophers: Women's Perspectives on Philosophical Traditions*, Brighton: Harvester Press.

Grosz, Elizabeth (1989) *Sexual Subversions: Three French Feminists*, Sydney: Allen & Unwin.

Hammond, Michael, Howarth, Jane and Keat, Russell (1991) *Understanding Phenomenology*, Oxford: Basil Blackwell.

Haraway, Donna (1985) 'A manifesto for cyborgs: science, technology and socialist feminism in the 1980s', *Socialist Review* 15, 2: 65–107.

—— (1989) 'The biopolitics of postmodern bodies: determinations of self in immune system discourse', *Differences* 1, 1: 3–43.

Harvey, Irene E. (1986) *Derrida and the Economy of Différance*, Bloomington: Indiana University Press.

Hegel, G.W.F. (1967) *Philosophy of Right*, trans. T.M. Knox, Oxford: Oxford University Press (this translation originally published by Clarendon Press, 1952).

—— (1971) *Philosophy of Mind* (Part Three of the *Encyclopaedia of the Philosophical Sciences*), trans. William Wallace (originally published by Oxford University Press, 1894) with the additions, trans. A.V. Miller, Oxford: Clarendon Press.

—— (1975) *Logic* (Part One of the *Encyclopaedia of the Philosophical Sciences*), trans. William Wallace, Foreword J.N. Findlay, Oxford: Oxford University Press.

—— (1977) *Phenomenology of Spirit*, trans. A.V. Miller, Foreword J.N. Findlay, Oxford: Oxford University Press.

Heidegger, Martin (1962) *Being and Time*, trans. John Macquarrie and Edward Robinson, New York: Harper & Row.

—— (1982) *The Basic Problems of Phenomenology*, trans. Albert Hofstadter, Bloomington: Indiana University Press.

—— (1984) *The Metaphysical Foundations of Logic*, trans. Michael Heim, Bloomington: Indiana University Press.

Hodge, Joanna (1987) 'Women and the Hegelian state', in Ellen Kennedy and Susan Mendus (eds) *Women in Western Political Philosophy: Kant to Nietzsche*, Brighton, Sussex: Harvester Press.

Hyppolite, Jean (1974) *Genesis and Structure of Hegel's Phenomenology of Spirit*, trans. Samuel Cherniak and John Heckman, Evanston: Northwestern University Press.

Irigaray, Luce (1985) *This Sex Which is not One*, trans. Catherine Porter with Carolyn Burke, Ithaca, New York: Cornell University Press.

—— (1987) *Speculum of the Other Woman*, trans. Gillian C. Gill, Ithaca, New York: Cornell University Press.

—— (1991) 'Sexual Difference' in Margaret Whitford (ed.) *The Irigaray Reader*, Oxford: Basil Blackwell (extract from *Ethique de la Différence Sexuelle*, Paris: Minuit, 1984).

Jardine, Alice (1985) *Gynesis: Configurations of Woman and Modernity*, Ithaca, New York: Cornell University Press.

Kirby, Vicki (1991) '*Corpus Delicti*: the body at the scene of writing', in Rosalyn Diprose and Robyn Ferrell (eds), *Cartographies: Poststructuralism and the Mapping of Bodies and Spaces*, Sydney: Allen & Unwin.

Kittay, Eva Feder and Meyers, Diana T. (eds) (1987) *Women and Moral Theory*, New Jersey: Rowman and Littlefield.

Kohlberg, Lawrence (1971) 'From is to ought: how to commit the naturalistic fallacy and get away with it in the study of moral development' in Theodore Mischel (ed.) *Cognitive Development and Epistemology*, New York: Academic Press.

Kojève, Alexandre (1980) *Introduction to the Reading of Hegel: Lectures on the Phenomenology of Spirit*, assembled by Raymond Queneau, ed. Allan Bloom, trans. James H. Nichols, Jr, Ithaca and London: Cornell University Press.

Krell, David Farrell and Wood, David (eds) (1988) *Exceedingly Nietzsche: Aspects of Contemporary Nietzsche-Interpretation*, London and New York: Routledge.

Leder, Drew (1990) *The Absent Body*, Chicago: University of Chicago Press.

Levinas, Emmanuel (1969) *Totality and Infinity*, trans. Alphonso Lingis, Pittsburgh: Duquesne University Press.

—— (1985) *Ethics and Infinity: Conversations with Philippe Nemo*, trans. Richard A. Cohen, Pittsburgh: Duquesne University Press.

Lloyd, Genevieve (1984) *The Man of Reason*, London: Methuen.

—— (1986) 'Selfhood, war and masculinity', in Carole Pateman and Elizabeth Gross (eds) *Feminist Challenges: Social and Political Theory*, Sydney: Allen & Unwin.

Locke, John (1967) *Two Treatises of Government*, 2nd edn, intro. Peter Laslett, Cambridge: Cambridge University Press.

—— (1975) *An Essay Concerning Human Understanding*, ed. Peter H. Nidditch, Oxford: Oxford University Press.

Macintyre, Alasdair (1985) *After Virtue: A Study in Moral Theory*, 2nd edn, London: Duckworth.

Mackenzie, Catriona (1986) 'Simone de Beauvoir: philosophy and/or the female body', in Carole Pateman and Elizabeth Gross (eds) *Feminist Challenges: Social and Political Theory*, Sydney: Allen & Unwin.

—— (1992) 'Abortion and embodiment', *Australasian Journal of Philosophy* 70, 2, June: 136–55.

McCumber, John (1988) 'Is a post-Hegelian ethics possible?', *Research in Phenomenology* 18: 125–47.

Martin, Emily (1990) 'Science and women's bodies: forms of anthropological knowledge', in Mary Jacobus, Evelyn Fox Keller, Sally Shuttleworth (eds) *Body/ Politics: Women and the Discourses of Science*, New York and London: Routledge.

Mauss, Marcel (1967) *The Gift: Forms and Functions of Exchange in Archaic Societies*, trans. Ian Cunnison, New York: Norton Library.

Merleau-Ponty, Maurice (1962), *Phenomenology of Perception*, trans. Colin Smith, London: Routledge & Kegan Paul.

—— (1964) *The Primacy of Perception and Other Essays*, ed. J.M. Edie, Evanston: Northwestern University Press.

Murphy, John M. and Gilligan, Carol (1980) 'Moral development in late adolescence and adulthood: a critique and reconstruction of Kohlberg's theory', *Human Development* 23: 159–70.

National Bioethics Consultative Committee (NBCC) (1990) *Surrogacy*, Report 1, Commonwealth of Australia, April.

Nietzsche, Friedrich (1967) *The Will to Power*, ed. and trans. Walter Kaufmann, trans. R.J. Hollingdale, New York: Random House.

—— (1968) *Twilight of the Idols and The Anti-Christ*, trans. R.J. Hollingdale, Harmondsworth, Middlesex: Penguin.

—— (1969) *On the Genealogy of Morals and Ecce Homo*, trans. Walter Kaufmann, New York: Random House.

—— (1972) *Beyond Good and Evil*, trans. R.J. Hollingdale, Harmondsworth, Middlesex: Penguin.

—— (1974) *The Gay Science*, trans. Walter Kaufmann, New York: Random House.

—— (1978) *Thus Spoke Zarathustra*, trans. Walter Kaufmann, Harmondsworth, Middlesex: Penguin.

—— (1979) 'On truth and lies in a nonmoral sense', *Philosophy and Truth: Selections from Nietzsche's Notebooks of the Early 1870s* ed. and trans. Daniel Breazeale, New Jersey: Humanities Press.

—— (1983) 'On the uses and disadvantages of history for life', *Untimely Meditations*, trans. R.J. Hollingdale, intro. J.P. Stern, Cambridge: Cambridge University Press.

—— (1983) 'Schopenhauer as educator', *Untimely Meditations*, trans. R.J. Hollingdale, intro. J.P. Stern, Cambridge: Cambridge University Press.

—— (1984) *Human, All Too Human: A Book for Free Spirits*, trans. Marion Faber with Stephen Lehmann, Lincoln: University of Nebraska Press.

Owens, Craig (1985) 'The discourse of others: feminists and postmodernism', in Hal Foster (ed.) *Postmodern Culture*, London: Pluto Press (originally published as *The Anti-Aesthetic*, Bay Press, 1983).

Parsons, Susan F. (1987) 'Feminism and the logic of morality: a consideration of the alternatives', *Radical Philosophy* 47: 2–12.

Pateman, Carole (1988) *The Sexual Contract*, Cambridge: Polity Press.

Patton, Paul (1989) 'Taylor and Foucault on power and freedom', *Political Studies* 37, 2: 260–76

—— (1991a) 'Nietzsche and the body of the philosopher', in Rosalyn Diprose and Robyn Ferrell (eds) *Cartographies: Poststructuralism and the Mapping of Bodies and Spaces*, Sydney: Allen & Unwin.

—— (1991b) 'Postmodern subjectivity: the problem of the actor (Zarathustra and the Butler)', in A. Yeatman (ed.) *Postmodern Critical Theorising*, special issue of *Social Analysis* 30, December.

Plato (1951) *The Symposium*, trans. Walter Hamilton, Harmondsworth, Middlesex: Penguin.

—— (1973) *Phaedrus*, published with *Letters VII and VIII*, trans. Walter Hamilton, Harmondsworth, Middlesex: Penguin.

Poole, Ross (1991) *Morality and Modernity*, London and New York: Routledge.

Porter, Elizabeth J. (1991) *Women and Moral Identity*, Sydney: Allen & Unwin.

Rich, Adrienne (1980) *On Lies, Secrets and Silence: Selected Prose 1966–1978*, London: Virago Press.

Sacks, Oliver (1985) *The Man who Mistook his Wife for a Hat*, London: Picador.

Sartre, Jean-Paul (1966) *Being and Nothingness: A Phenomenological Essay on Ontology*, trans. Hazel E. Barnes, New York: Pocket Books.

Sawicki, Jana (1986) 'Foucault and feminism: towards a politics of difference', *Hypatia* 1, 2: 23–36.

—— (1988) 'Identity politics and sexual freedom: Foucault and feminism' in Irene Diamond and Lee Quinby (eds) *Foucault and Feminism: Reflections on Resistance*, Boston: Northeastern University Press.

Scaltsas, Patricia Ward (1992) 'Do feminist ethics counter feminist aims?', in Eve Browning Cole and Susan Coultrap-McQuinn (eds) *Explorations in Feminist Ethics: Theory and Practice*, Bloomington: Indiana University Press.

Schenck, David (1986) 'The texture of embodiment: foundation for medical ethics', *Human Studies* 9, 1: 43–54.

Scott, Charles E. (1988) 'Heidegger and the question of ethics', *Research in Phenomenology* 18: 23–40.

—— (1990) *The Question of Ethics: Nietzsche, Foucault, Heidegger*, Bloomington: Indiana University Press.

Sherwin, Susan (1989) 'Feminist and medical ethics: two different approaches to contextual ethics', *Hypatia* 4, 2: 57–72.

Sissa, Giulia (1989) 'Subtle Bodies', in Michel Feher (ed.) with Romana Naddaff and Nadia Tazi *Fragments of a History of the Human Body, Part III*, New York: Zone.

Sophocles (1960) *Antigone*, trans. Peter Arnott, Arlington, Illinois: AHM Publishing Corp.

Spivak, Gayatri Chakravorty (1982) 'Displacement and the discourse of woman', in Mark Krupnick (ed.) *Displacement, Derrida and After*, Bloomington: Indiana University Press.

—— (1984) 'Love me, Love my Ombre, Elle', *Diacritics* 14, 4, Winter: 19–36.

—— (1989) 'Feminism and deconstruction, again: negotiating an unacknowledged masculinism' in Teresa Brennan (ed.) *Between Feminism and Psychoanalysis*, London: Routledge.

Stone, Sandy (1991) 'The *empire* strikes back: a posttranssexual manifesto', in Julia Epstein and Kristina Straub (eds) *Body Guards: The Cultural Politics of Gender Ambiguity*, New York and London: Routledge.

Vasseleu, Cathryn (1991) 'Life itself' in Rosalyn Diprose and Robyn Ferrell (eds) *Cartographies: Poststructuralism and the Mapping of Bodies and Spaces*, Sydney: Allen & Unwin.

Warren, Mark (1985) 'Nietzsche and political philosophy', *Political Theory* 13, 2: 183–212.

Whitford, Margaret (1991) *Luce Irigaray: Philosophy in the Feminine*, London and New York: Routledge.
Wood, Allen W. (1990) *Hegel's Ethical Thought*, Cambridge: Cambridge University Press.
Wood, David (1988) 'Nietzsche's transvaluation of time', in David Farrell Krell and David Wood (eds) *Exceedingly Nietzsche: Aspects of Contemporary Nietzsche-Interpretation*, London: Routledge.
Young, Iris Marion (1990a) *Throwing Like a Girl and Other Essays in Feminist Philosophy and Social Theory*, Bloomington: Indiana University Press.
—— (1990b) *Justice and the Politics of Difference*, Princeton: Princeton University Press.
Young, Robert (1986) *Personal Autonomy: Beyond Negative and Positive Freedom*, New York: St Martin's Press.

Index

